GRAPHING POWER

High School Activities for the TI-81 and TI-82

The Graphing Technology in Mathematics Project

Dale Seymour Publications

Photo TI-81 and TI-82 reprinted courtesy of Texas Instruments, Inc.
All rights reserved.

Project Editor: Joan Gideon
Production Coordinator: Claire Flaherty
Text and Cover Design: Lisa Raine
Cover Image: Thomas Lochran, Image Bank

Published by Dale Seymour Publications, an imprint of the Alternative
Publishing Group of Addison-Wesley Publishing Company.

Order Number DS25301

ISBN 0-86651-827-4

4 5 6 7 8 9 10-ML-98 97

DALE
SEYMOUR
PUBLICATIONS
P.O. BOX 10888
PALO ALTO, CA 94303

The Graphing Technology in Mathematics Project

Sponsored by

Trinity University, The San Antonio Independent School District, and The North East Independent School District, San Antonio, Texas

This Project was made possible by a grant from the Dwight D. Eisenhower Mathematics and Science Programs Texas Higher Education Coordinating Board.

Project Director

David D. Molina, Ph.D., Trinity University

Project Co-Directors

Mary Ann Sample, San Antonio Independent School District
Lu Ann Weynand, North East Independent School District

Graduate Assistants

Wallace Ross, Trinity University
Stacy Dill, Trinity University
Karen Stenson, Trinity University

Editor

Alma Aguirre, Thomas Jefferson High School

Contributing Editor

Laurie Bergner, Robert E. Lee High School

Participants and Contributing Writers

Shelia Lloyd
Brackenridge HS

Linda Patterson
Brackenridge HS

Janet Starkey
Brackenridge HS

Sylvia Carreón
Burbank HS

Ascencion Gonzalez
Burbank HS

Yolanda Hughes
Burbank HS

Michael Russell
Burbank HS

Flor de Maria
Wukovits
Burbank HS

Roland Rios
Churchill HS

Nancy Cogdell
Connell MS

Debbie Fisher
Connell MS

Susan Tustison
Connell MS

Evelyn Williams
Cooper MS

Diana Harwood
Driscoll MS

Mike Hansen
Eisenhower MS

Vera Hayes
Eisenhower MS

Maria E. Cantú
Edison HS

Magda Gonzales
Edison HS

Marie Graham
Edison HS

Susan Ng
Edison HS

Diana Hughes
Garner MS

Rebecca Moch
Harris MS

Kim Rodriguez
Harris MS

George Watkins
Healy Murphy

Gary Frank Clark
Highlands HS

Shirley Rich
Highlands HS

Naida Brueland
Irving MS

Anna Pilgrim
Irving MS

Karen Allbright
Jackson MS

Lee Anne Leister
Jackson MS

Alma Aguirre
Jefferson HS

Cathy Morgan
Judson HS

Annette Plowman
Judson HS

Rosie Sabala
King MS

Ana Thompson
Krueger MS

Valorie Fayfich
Lanier HS

Carole Gould
Lee HS

Shan Schanding
Lee HS

Paul Tisdel
Lee HS

Anthony Plummer
Lee HS

Juanita B. Martinez
Lowell MS

Maria T. Muñoz
Lowell MS

Michael Maltrud
Marshall HS

Andy Davis
Navarro AC

David Monteith
Nimitz MS

Joe Rosseett
Page MS

June Draper
Rogers MS

Sergio Perales
Roosevelt HS

William Lauderdale
Sam Houston HS

James F. (Skip) Lee
Sam Houston HS

Karla Escalera
St. Gregory

Stella Vasquez
St. Gregory

Patricia Gomez
Tafolla MS

Betsy Burke
Trinity Univ.

Erin Roberson
Trinity Univ.

Shannon Walsh
Trinity Univ.

Leslie Cate
Twain MS

Rosemary Hidalgo
Twain MS

Andrew Stepanski
Twain MS

Rudy Tamez
Twain MS

Evangeline Tovar
Wheatley MS

Sandy Rothrock
White MS

Mary Lynne
Montgomery
White MS

Peter Barrera
Whittier MS

Preface

The graphing calculator gives students the graphing power to explore mathematics concepts. With it students can see the slope of a line change as they key in different coefficients of x; they can watch the phase shift and the amplitude vary as they alter the coefficients in trigonometric functions. By enabling students to quickly generate examples, the graphing calculator allows them to see patterns, discover relationships, and validate conjectures. Educators are advocating use of this powerful tool in mathematics classrooms.

For two summers, groups of teachers gathered at Trinity University in San Antonio, Texas, to study the use of the graphing calculator in the middle school and high school. Participants generated lessons they could use in their classrooms. The two books in the *Graphing Power* series represent the results of their efforts. The work done by these teachers will help you make effective use of graphing technology in your classroom. These activities are suggestions and models for lessons that make the best use of the calculator; they will give you ideas for developing your own lessons. Use them in conjunction with concrete materials, other technologies, and traditional lessons.

These activities are intended for use in a classroom where students are encouraged to explore and discover, preferably in groups of two to four. Instructions in the lessons provide help using the TI-81 and TI-82, but some experience with the graphing calculator is assumed. Have calculator manuals available for reference as you work through the lesson before presenting it to your class and as your students complete the activity.

If you need some activities based on simpler mathematics for pre-algebra or general math classes, you can find more activities in *Graphing Power: Middle School Activities for the TI-81 and TI-82*.

Contents

TOPIC 1

Working with Numbers

Order of Operations

Level   Pre-Algebra, Algebra I

Using "Operations Ordered" students will

- Determine how the use of parentheses changes the order of operations
- Place parentheses to produce a desired result
- Describe in writing what the order of operations is
- Generate examples, solve, and provide a written explanation of solutions

Teaching the Lesson

- Most students will be able to complete this lesson without difficulty.
- Check students' work often, to make sure that they are drawing the correct conclusions about the order of operations.

Solutions

11. a. $24 \div (3 + 5) \cdot 2 = 6$

   b. $3 \cdot (4 - 2) \cdot 3 = 18$

   c. $(24 - 4) \cdot 2 = 40$

   d. $(12 - 2^2) \div 4 = 2$

   e. $4 \cdot 6 + 3 \cdot 3 = 33$

12. $9 - (3 \cdot 4 + 5 \cdot 13) \div (9 + 4 \div 2) = 2$

Operations Ordered

1. Use your calculator to find $3 + 4 \times 5$. Which operation, $3 + 4$ or 4×5, was computed first by the calculator? Enter $3 + (4 \times 5)$ and $(3 + 4) \times 5$ to verify your answer.

2. Work these problems on your calculator and answer the questions that follow.

   a. $6 + 2 \times 8 =$ d. $8 + 9 \div 3 =$

   b. $6 + (2 \times 8) =$ e. $8 + (9 \div 3) =$

   c. $(6 + 2) \times 8 =$ f. $(8 + 9) \div 3 =$

3. Did your answer change when the multiplication or division was grouped in parentheses?

4. Did your answer change when the addition was grouped in parentheses?

5. When parentheses were not used, what operation was done first?

6. When the parentheses were used, what operation was done first?

7. Work the following problems on your calculator and answer the question below.

   a. $\dfrac{2 + 3}{6 + 4} =$ c. $\dfrac{(12 + 6)}{(6 + 3)} =$

   b. $\dfrac{(2 + 3)}{(6 + 4)} =$ d. $\dfrac{(12 + 6)}{(6 + 3)} =$

8. Why are the answers to 7a and 7b different? Why are the answers to 7c and 7d different? Why are parentheses important when working with fractions?

9. Generate two examples of your own similar to those in question 7 and write a sentence explaining what is happening and why the answers are different.

10. Work the following problems on paper, then verify your answers on the calculator.

    a. $9 + 45 \div 9 \cdot 8 =$

    b. $(53 - 8) \div 5 =$

    c. $8 + 32 \div 4 \cdot 5 =$

    d. $\dfrac{10 \cdot 4}{14 - 9} =$

    e. $9^2 \cdot 3 \cdot 5 =$

    f. $11 + 3 (19 + 2) =$

    g. $14 + (3^3 - 7) =$

11. Classify each equation as true or false, and if false, place parentheses where needed to make the equation true.

    a. $24 \div 3 + 5 \cdot 2 = 6$ d. $12 - 2^2 \div 4 = 2$

    b. $3 \cdot 4 - 2 \cdot 3 = 18$ e. $4 \cdot 6 + 3 \cdot 3 = 33$

    c. $24 - 4 \cdot 2 = 40$

12. Use the calculator and its EDIT features to make the following equation true.

$$9 - 3 \cdot 4 + 5 \cdot 13 \div 9 + 4 \div 2 = 2$$

13. Use the calculator and the digits from the year of your birth to create an expression equal to each of the numbers 0 through 10. The digits can be used in any order, but can't be repeated. All operations may be used and parentheses may be inserted when needed.

For example, 1973

$$1 \cdot 9 + 3 - 7 = 5$$

$$\frac{9 - 3}{7 - 1} = 1$$

$$(9 - 3) + 1 - 7 = 0$$

Graphing Power © Dale Seymour Publications

Exponents

Level   Pre-Algebra, Algebra I

Using "Powers" students will

- Evaluate numerical expressions with exponents and compare results
- Use relational testing on the graphing calculator to determine the truth of given statements

Teaching the Lesson

- The $\boxed{\text{TEST}}$ menu results in one of two answers, 0 for false and 1 for true.
- Careful placement of parentheses is important.

Solutions

1. a and d are equivalent
   c, e, and f are equivalent

2. a and f are equivalent
   b, c, e, and h are equivalent

3. a and c are equivalent

Powers

1. Predict which of these expressions are equivalent and state the rule that supports the prediction.

   Use the calculator to check your predictions

   a. $(4^2)(4^3)$ =   f. $(4^3)^2$ =

   b. $4^2 + 4^3$ =   g. $2 \cdot 4^5$ =

   c. $(4^2)^3$ =   h. $4 \cdot 6$ =

   d. 4^5 =   i. $4 \cdot 5$ =

   e. 4^6 =

2. Predict which of these expressions are equivalent and state the rule that supports the prediction.

   Use the calculator to check your predictions

   a. $(3 \cdot 2^3)^2$ =   f. $3^2 (2^3)^2$ =

   b. $3^5 \cdot 2^5$ =   g. $3 \cdot (2^3)^2$ =

   c. $(3 \cdot 2)^5$ =   h. $(3^5 \cdot 2)(2^3 \cdot 2)$ =

   d. $(3 \cdot 2)^6$ =   i. $3 \cdot (2 \cdot 3)^2$ =

   e. $3^5 \cdot 2^3 \cdot 2^2$ =

3. Predict which of these expressions are equivalent and state the rule that supports the prediction.

   a. $2^3 (2^4)$ =   c. 2^7 =

   b. 2^{12} =   d. $2 \cdot 7$ =

Graphing Power © Dale Seymour Publications

4. Predict which statements are true. To test your prediction on the calculator, enter the expression from the left side of the equation. Use the $\boxed{\text{TEST}}$ menu, select =. Enter the second expression, then $\boxed{\text{ENTER}}$. If the calculator returns a 1, the test is *true*. If the calculator returns a 0, the result is *false*.

Statement	Prediction	Result	Explanation
$(-5)^3 = 5^3$			
$(-5)^3 = -5^3$			
$(-5)^4 = -5^4$			
$((-1) \cdot 5)^3 = (-5)^3$			
$(-5) \cdot (-5) \cdot (-5) = -5^3$			
$(-3)^4 = 3^4$			
$-3^4 = (-3)^4$			
$-(3)^4 = -3 \cdot -3 \cdot -3 \cdot -3$			

5. Describe how to find the product of the expression $a^3 \cdot a^4$. Make a problem of your own that demonstrates you know this rule.

6. Describe how to find the value of the expression $(a^3)^4$. Make a problem of your own that demonstrates you know this rule.

7. Describe how to find the value of the expression $(a \cdot b)^3$. Make a problem of your own that demonstrates you know this rule.

Negative Exponents

Level   Pre-Algebra, Algebra I, Algebra II

Using "Negative Exponents" students will

- Graph exponential functions

- Record values from the graph

- Make conjectures regarding both positive and negative exponents

Teaching the Lesson

- When using TRACE to find the values for 2^x, the x value on the viewing screen is the exponent x. That is $x = -0.1$ $y = 0.93308\ldots$ represents $2^{-0.1} = 0.93308\ldots$.

- The shaded boxes in the chart do not require an answer. Completing those boxes could be an extension of the lesson.

- Scrolling left and right on the screen and changing the RANGE will enable the students to make better conjectures.

Negative Exponents

	TI-81	**TI-82**
Suggested Range	$X_{min} = -4.8$	$X_{min} = -4.8$
	$X_{max} = -4.7$	$X_{max} = 4.6$
	$X_{scl} = 1$	$X_{scl} = 1$
	$Y_{min} = -6.4$	$Y_{min} = -6.4$
	$Y_{max} = 6.4$	$Y_{max} = -6.2$
	$Y_{scl} = 1$	$Y_{scl} = 1$
	$X_{res} = 1$	

1. Enter $Y_1 = 2^x$ and $\boxed{\text{GRAPH}}$

2. Follow the directions below to complete the chart.
   a. Use $\boxed{\text{TRACE}}$ to complete the second column of the table.
   b. Convert all decimals to fractions to complete column 3.
   c. Rewrite each fraction so that the denominator is a power of 2.

3. What relationship exists between the first and last columns of the chart?

4. Write a rule to express this relationship.

5. What happens to y as x gets larger?

6. What happens to y as x gets smaller?

7. When will y equal zero?

2^x	$\boxed{\text{TRACE}}$ Value	Fraction Equivalent	$\dfrac{1}{2^{\square}}$ Form
2^5			
2^4			
2^3			
2^2			
2^1			
2^0			
2^{-1}			$\dfrac{1}{2^{\square}}$
2^{-2}			$\dfrac{1}{2^{\square}}$
2^{-3}			$\dfrac{1}{2^{\square}}$
2^{-4}			$\dfrac{1}{2^{\square}}$
2^{-5}			$\dfrac{1}{2^{\square}}$

8. When will y have a negative value?

9. Complete the table for $y = 3^x$. Answer the questions that follow.

10. What relationship exists between the first and last columns of the chart?

3^x	TRACE Value	Fraction Equivalent	$\frac{1}{3^\square}$ Form
3^5			
3^4			
3^3			
3^2			
3^1			
3^0			
3^{-1}			$\frac{1}{3^\square}$
3^{-2}			$\frac{1}{3^\square}$
3^{-3}			$\frac{1}{3^\square}$
3^{-4}			$\frac{1}{3^\square}$
3^{-5}			$\frac{1}{3^\square}$

11. Write a rule to express this relationship.

12. What happens to y as x gets larger?

13. What happens to y as x gets smaller?

14. When will y equal zero?

15. When will y have a negative value?

Graphing Power © Dale Seymour Publications

16. Complete a similar table for $y = -(2^x)$. What effect does this have on the function? Compare the graph of this equation with the graph of $y = 2x$. List the similarities and differences in both equations.

$-(2^x)$	TRACE Value	Fraction Equivalent	$\frac{1}{2^\square}$ Form
$-(2^5)$			
$-(2^4)$			
$-(2^3)$			
$-(2^2)$			
$-(2^1)$			
$-(2^0)$			
$-(2^{-1})$			$\frac{1}{2^\square}$
$-(2^{-2})$			$\frac{1}{2^\square}$
$-(2^{-3})$			$\frac{1}{2^\square}$
$-(2^{-4})$			$\frac{1}{2^\square}$
$-(2^{-5})$			$\frac{1}{2^\square}$

Challenge

- Complete a similar table for $y = (-2)^x$. What effect does the negative base have on the function? Does a negative affect the graph the way a negative exponent does? Explain.

TOPIC 2

Working with Patterns

Fibonacci Sequence

Level   Algebra II, Geometry

Using "Fibonacci Numbers and the Golden Ratio" the students will

- Make a scatter plot with the Fibonacci numbers along the x-axis and the ratio of consecutive Fibonacci numbers along the y-axis

- Enter and run a program to observe the connection between the Fibonacci sequence, the golden ratio, and its scatter plot

- Draw conclusions regarding the relationship between the golden ratio and Fibonacci sequence

Teaching the Lesson

- Background information on Fibonacci numbers and the golden ratio can be found in

  Trudi Garland. *Fascinating Fibonaccis*. Palo Alto, Ca.: Dale Seymour Publications, 1987.

  Gath Runion. *The Golden Section*. Palo Alto, Ca.: Dale Seymour Publications, 1990.

- Students may notice that the display of the ratio remains unchanged prior to the end of the program. The calculator does not display all the decimal places it considers.

- Students' displays will stop changing after a while even though the program needs to be run further.

- This program takes some time to enter.

Command Locations for the Golden Ratio Program

Command	Where to Find It
the little arrow	STO key
ClrStat	STAT menu, DATA submenu, ClrStat [TI-81]
	STAT menu, ClrList option [TI-82]
Lbl	PRGM menu, EDIT submenu
ClrHome	PRGM menu, I/O submenu
Disp	PRGM menu, I/O submenu
to get quote marks	ALPHA , (plus key)
to get a space	ALPHA , (zero key)
If	PRGM menu, CTL submenu

=	$\boxed{\text{TEST}}$ menu
Goto	$\boxed{\text{PRGM}}$ menu, CTL submenu
Pause	$\boxed{\text{PRGM}}$ menu, CTL submenu
ClrDraw	$\boxed{\text{DRAW}}$ menu
Scatter	$\boxed{\text{STAT}}$ menu, DRAW submenu [TI-81]
	$\boxed{\text{STAT PLOT}}$, $\boxed{\text{ENTER}}$, $\boxed{\text{ON}}$, choose scatter from type if using the [TI-82]

Solutions

6. The ratio of consecutive Fibonacci numbers approaches the value of the golden ratio—approximately 1.618033.

Graphing Power © Dale Seymour Publications

Fibonacci Numbers and the Golden Ratio

Fibonacci sequence $1, 1, 2, 3, 5, 8, 13, \ldots, n-1, n, (n-1) + n, \ldots$

Golden Ratio $\dfrac{1 + \sqrt{5}}{2}$

1. Using consecutive pairs of elements of the Fibonacci sequence, complete the following chart. Use the second number of the pair for x, let y be the ratio of the first number in the pair to the second. Extend the chart if necessary.

x Fibonacci Number	Previous Fibonacci Number	y Fib. Num. (x) Previous Fib. Num.	Point (x, y)
1	1	1	(1, 1)
2	1	2	(2, 2)
3	2	1.5	(3, 1.5)

2. Plot the points for the chart above on the graph provided.

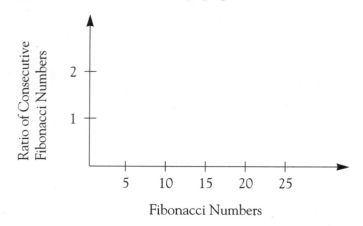

3. As the value of x increases, what do you notice about the value of y?

Graphing Power © Dale Seymour Publications

4. The following program displays consecutive pairs of numbers from the Fibonacci sequence. It also displays the ratio of the consecutive numbers and asks if that value is the golden ratio.

Suggested Range

$X_{min} = 0$
$X_{max} = 55$
$X_{scl} = 5$
$Y_{min} = 0$
$Y_{max} = 2.4$
$Y_{scl} = 1$
$X_{res} = 1$

TI-81

```
:(√(5)+1)/2→R
:0→A
:1→B
:ClrStat
:0→N
:Lbl T
:ClrHome
:A+B→C
:C/B→G
:N+1→N
:C→{x}(N)
:G→{y}(N)
:Disp "FIBONACCI NOS."
:Disp B
:Disp C
:Disp " "
:Disp "GOLDEN RATIO?"
:Disp G
:If G=R
:Goto E
:Disp "NOT YET"
:Pause
:B→A
:C→B
:Goto T
:Lbl E
:Disp " "
:Disp "BEST APPROX ON THE TI-81"
:Pause
:ClrDraw
:Scatter
```

TI-82

```
:ClrList L₁, L₂
:(√(5)+1)/2→R
:0→A
:1→B
:0→N
:Lbl T
:ClrHome
:A+B→C
:C/B→G
:N+1→N
:C→L₁(N)
:G→L₂(N)
:Disp "FIBONACCI NOS."
:Disp B
:Disp C
:Disp " "
:Disp "GOLDEN RATIO?"
:Disp G
:If G=R
:Goto E
:Disp "NOT YET"
:Pause
:B→A
:C→B
:Goto T
:Lbl E
:Disp " "
:Disp "BEST APPROX ON THE TI-82"
:Pause
:ClrDraw
:Plot 1 (Scatter, L₁, L₂)
:Disp Graph
```

Press ENTER each time to see the next screen. Your final screen will be a graph of the points.

5. Does the graph at the end of the program look familiar? What is different about it? (Look at the Range.)

6. Explain the relationship between the ratio of consecutive Fibonacci numbers and the golden ratio.

7. Will the ratio of consecutive Fibonacci numbers ever equal the golden ratio? Explain your answer.

Challenge

Research and report on the Fibonacci sequence, the golden ratio, or Leonardo Fibonacci.

Graphing Power © Dale Seymour Publications

Patterns in Quadratic Equations

Level   Pre-Algebra, Algebra I

Using "Graphing Statistics" students will

- Enter data in the $\boxed{\text{STAT}}$ mode
- Discover the solution to real world problems
- Interpret data

Teaching the Lesson

- Make sure that both the $\boxed{\text{STAT}}$ and $\boxed{\text{DRAW}}$ menus are cleared before starting this lesson.
- The procedures in the Analysis section of this lesson might be difficult for some students to follow independently. Working in groups or as a class might be more successful.
- The key strokes for the TI-81 and TI-82 are significantly different in the Analysis section and are given for both calculators on the student pages.
- Questions 11–21 take an in-depth look at the function and could be used as an extension of the first part of this lesson.

Solutions

3. $57.60

4. 24 people

5. 12 price increases

6. $2.40

Graphing Statistics

Clear the STAT and DRAW menus.

Suggested Range

$X_{min} = -10$
$X_{max} = 85$
$X_{scl} = 10$
$Y_{min} = -10$
$Y_{max} = 60$
$Y_{scl} = 10$
$X_{res} = 1$

The Problem

The Oak Garden swimming pool charges $1.20 per day for members to swim on weekdays. At this price, approximately 36 people come to swim on a typical day. Millie Waters, the manager, has discovered that, for every 10¢ that she increases the price, one less person comes to swim.

Your job is to discover what price Millie should charge for a day of swimming in order to earn the *maximum income*.

The Activity

1. Before you begin, estimate how much you think Millie should charge for each person to swim on weekdays in order to earn the maximum income.

    Estimate _____ per person

    Explain how you arrived at this estimate.

Graphing Power © Dale Seymour Publications

2. Fill in the missing data to complete the chart. Follow the pattern to get the next entry.

Number of Price Increases	Admission Price, x	Number of Swimmers	Total Income
0	$1.20	36	$43.20
1	$1.30	35	$45.50
2	$1.40		
3			
4			
5			
6			
7			
8			
9			
10			
11			
12			
13			
14			
15			
16			
17			
18			
19			
20			
21			
22			
23			
24			

Graphing Power © Dale Seymour Publications

Summarize the Data

3. What is the pool's maximum daily income?

4. How many people come to swim when the pool is earning this amount?

5. How many 10¢ price increases does it take to get the maximum income?

6. What is the price of admission when the pool earns the maximum income?

7. Compare this amount to your estimate.

Analysis

On the TI-81

- Access the [STAT] menu. Move the cursor to the DATA submenu and select the EDIT option. Use the data from your chart to enter the x and y values. The x values are the "number of swimmers," and the y values are the "total incomes."

- When all the data has been entered, go to the [STAT] menu and choose the [DRAW] submenu, select xyLine, press [ENTER]. This graphs and connects the (x, y) points.

Graphing Power © Dale Seymour Publications

On the TI-82

- Access the [STAT] menu. Select edit in the EDIT submenu. Enter "number of swimmers" in L_1 and enter "total incomes" in L_2.

- Access the [STAT PLOT] menu, choose Plot 1; turn ON; from Type, select scatter plot; press [GRAPH].

8. What does the x-axis represent?

9. What does the y-axis represent?

   [TRACE] does not work in this mode, but you can use the arrow keys to explore the graph.

10. Explore the graph and then write a short analysis of what you find.

    The graph of this data is part of a quadratic function. The legs that curve down continue to do so. There is only one number of swimmers that gives the maximum income. It is shown at the vertex (the top point of the graph).

    With the graph on the screen, access the [DRAW] menu. Choose the Line option. Your graph will now be back on screen. Position the cursor to the left endpoint of the graph, press [ENTER] and draw a line (using the down and left arrows) until you have a line that looks like an extension of the graph, press [ENTER] to complete the drawing. Use the right arrow to complete the other leg of the graph using the same procedure. When you have the graph completed (both legs extending down to the x-axis) answer the following in terms of swimmers and income.

11. What number of swimmers gives the maximum income?

12. Use the arrow keys to move the cursor to the point where the right leg of the graph intersects the x-axis. What do the x, y values represent at this point?

13. If you completed the graph properly the left leg of the graph intersects the y-axis. Move the cursor to this intersection. What do the x, y values represent at this point?

14. Put the cursor on the point where the left leg of the graph intersects the x-axis. What do the x, y values represent at this point?

15. What is the expression that represents the number of swimmers?

16. What is the expression that represents the total income at the pool?

17. Why does y represent the income and x represent the number of swimmers? Could we use x to represent the income and y the number of swimmers?

18. If income doesn't matter, at what price would the most swimmers come to the pool?

19. How many swimmers will come at that price?

20. How would the answer change if, at the beginning of the problem, Millie had discovered that two fewer swimmers would come each time she raised the price by ten cents?

21. How would the answer change if Millie had discovered that three fewer swimmers would come each time she raised the price by 25 cents?

Graphing Power © Dale Seymour Publications

TOPIC 3

Equations and Inequalities

Solving Equations

Level   Pre-Algebra, Algebra I

Using "Find the Intersection" students will

- Enter the expressions for each side of an equation in the $\boxed{\text{Y=}}$ menu
- Graph the expressions
- Find the point of intersection thus solving the equation
- Verify the solution by substitution on the calculator
- Use the $\boxed{\text{Y-VARS}}$ menu to verify the solution

Teaching the Lesson

- Clear all entries in the $\boxed{\text{Y=}}$ menu before starting this lesson.
- To check solutions, the cursor must be on the point of intersection as the calculator will be using that point of intersection to verify the solution.
- The $\boxed{\text{Y-VARS}}$ menu contains the equations that are entered into the $\boxed{\text{Y=}}$ menu.
- The given $\boxed{\text{RANGE}}$ and $\boxed{\text{WINDOW}}$ values will give the students integer values when they $\boxed{\text{TRACE}}$ to find the solutions.

Find the Intersection

	TI-81	**TI-82**
Suggested Range	$X_{min} = -47$	$X_{min} = -47$
	$X_{max} = 48$	$X_{max} = 47$
	$X_{scl} = 1$	$X_{scl} = 1$
	$Y_{min} = -32$	$Y_{min} = -31$
	$Y_{max} = 31$	$Y_{max} = 31$
	$Y_{scl} = 2$	$Y_{scl} = 2$
	$X_{res} = 1$	

1. To solve the equation $x + 5 = 11$, enter $Y_1 = x + 5$ in the $\boxed{Y=}$ menu.

2. Enter $Y_2 = 11$, for the right side of the equation. Press $\boxed{\text{GRAPH}}$.

3. Use $\boxed{\text{TRACE}}$ and the arrow keys to explore the lines. Find the point of intersection for the two lines. List the ordered pair below.

   $(x, y) = ($ _____ , _____ $)$

4. How do you know this is the point of intersection?

5. Is this the *solution* of the equation? Explain.

   To check the solution, make sure that the cursor is on the intersection of the two lines then $\boxed{\text{QUIT}}$. Press $\boxed{\text{XIT}}$ then $\boxed{\text{ENTER}}$. This is the value of x that makes this equation true (the solution).

   Access the $\boxed{\text{Y-VARS}}$ menu, select Y_1, then $\boxed{\text{ENTER}}$. If using the TI-82, Y_1 is in the function submenu in $\boxed{\text{Y-VARS}}$.

6. Is this the same y value as your point of intersection?

7. Repeat for Y_2. Is this the same y value as your point of intersection?

Graphing Power © Dale Seymour Publications

8. Explain what is meant by a *solution* of an equation.

9. Explain how graphing can be used to find the solution to an equation.

10. How do you verify (check) the solution of an equation?

11. Solve these equations in the same manner by entering the left side of the equation in Y_1 and the right side of the equation in Y_2. Be sure to verify your solutions.

a. $x - 3 = -10$

b. $x - 4 = 6$

c. $x - 2 = -7$

d. $2x = 14$

e. $5x = -40$

f. $\frac{x}{2} = 9$

g. $\frac{x}{3} = -4$

h. $4x + 7 = 15$

i. $6x + 2 = -4$

j. $3x - 4 = 8$

k. $5x - 6 = -36$

l. $2x + 5 = 7x - 10$

m. $2(x + 3) = -3(x + 2)$

Graphing Power © Dale Seymour Publications

Absolute Value

Level   Algebra I, Algebra II

Using "Absolutely Valuable" students will

- Solve absolute value inequalities algebraically

- Use the graphing capabilities to verify solutions graphically and solve application problems

Teaching the Lesson

- To graph Y_1, Y_2, and Y_3 separately. Deselect the equations you do not want graphed by going to the Y= menu and moving the cursor on top of the darkened equals sign and press ENTER. The darkened square is replaced with a regular equal sign and the equation will not graph. To re-select the equation, repeat the above instructions. The darkened square will once again appear and the equation will be graphed.

- The ≥ and ≤ symbols are in the TEST menu.

Solutions

9. Answers will vary, however, there is a $2.50 profit per pizza. Total profit depends on the number of pizzas sold.

Absolutely Valuable

Solving Absolute Value Inequalities

Example 1 $| 2x | \leq 8$ means

$2x \geq -8$ and $2x \leq 8$

$x \geq -4$ and $x \leq 4$

For $x \geq -4$ and $x \leq 4$, we write $-4 \leq x \leq 4$ and take the *inter-section* of the two graphs. This is also known as a *conjunction*.

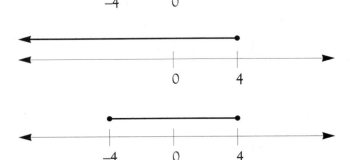

Example 2 $| 2x | \geq 8$ means

$2x \leq -8$ or $2x \geq 8$

$x \leq -4$ or $x \geq 4$

For $x \geq 4$ or $x \leq -4$, we take the *union* of the two graphs. This is also known as a *disjunction*.

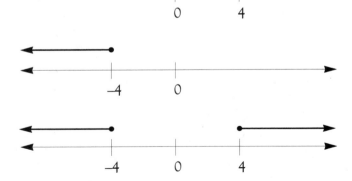

1. Compare the answers in examples 1 and 2. How are they alike? Why? How are they different? Why?

2. Solve algebraically and graph

   a. $| 5x | \leq 20$

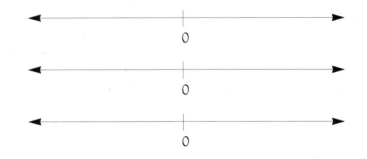

Graphing Power © Dale Seymour Publications

b. $|y - 3| > 12$

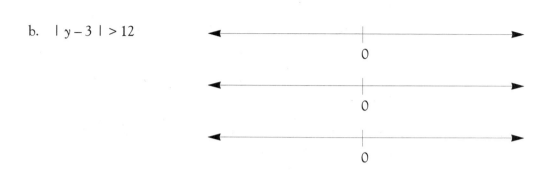

3. Describe the difference between each part of the solution and the final solution in question 2a and question 2b.

4. Solve $|2x| \leq 8$. Use the $\boxed{Y=}$ menu to graph the solution to the inequalities.

	TI-81	**TI-82**
Suggested Range	$X_{min} = -9.4$	$X_{min} = -9.4$
	$X_{max} = 9.6$	$X_{max} = 9.4$
	$X_{scl} = 1$	$X_{scl} = 1$
	$Y_{min} = -10$	$Y_{min} = -10$
	$Y_{max} = 10$	$Y_{max} = 10$
	$Y_{scl} = 1$	$Y_{scl} = 1$
	$X_{res} = 1$	

Enter the following in the $\boxed{Y=}$ menu:

$Y_1 = 2x \leq 8$

$Y_2 = 2x \geq -8$

$Y_3 = abs\ (2x) \leq 8.$

$\boxed{\text{GRAPH}}$ each one separately.

Solve $|2x| \geq 8$ the same way, adjusting the $\boxed{Y=}$ values accordingly.

Graph Y_3 and $\boxed{\text{TRACE}}$. What happens as you trace along the line? What happens when the cursor jumps off the line? What does this mean in regards to the solution?

5. Use the $\boxed{Y=}$ menu and $\boxed{TRACE}$ to check your solutions to question 2.

6. Compare and contrast the algebraic solution with the calculator's solution.

7. Solve and graph each inequality. Graph each portion of the solution, as well as the final solution. $\boxed{TRACE}$ to verify your solution.

a. $|z + 3| \leq 15$

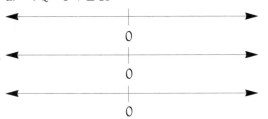

c. $|b - 4| < 12$

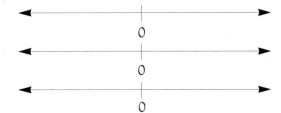

b. $|2x - 4| \geq 12$

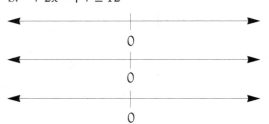

d. $2|8x + 6| \leq 24$

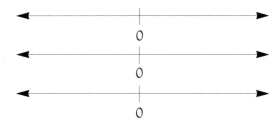

8. Explain the differences in the solutions to the absolute value inequalities for > and <.

9. A pizza company has established a mathematical model to represent their profit. This equation represents the region within which the company's sales have resulted in a profit: $|2p + 5| \leq 10$ where p represents the profit for each pizza sold. Find the profit margin for this company. Do you think this company is making large profits? Why or why not?

Graphing Power © Dale Seymour Publications

T O P I C 4

Statistics and Probability

Line of Best Fit

Level   Pre-Algebra, Algebra I, Algebra II

Using "Dis and Data" students will

- Study data and prepare scatter plots

- Observe differences and similarities among line graphs and scatter plots

- Draw a scatter plot on the graphing calculator

- Find the *line of best fit* for a set of data

- Explore the regression model capabilities of the graphing calculator by using them to analyze student-generated data

- Organize the acquired information from the student-generated data for a classroom presentation

Teaching the Lesson

- Clear the $\boxed{\text{STAT}}$ and $\boxed{\text{Y=}}$ menus prior to starting the lesson.

- The TI-82 contains the scatter plot options in the $\boxed{\substack{\text{STAT} \\ \text{PLOT}}}$ menu. The data entry will be entered into L_1 and L_2 in the $\boxed{\text{STAT}}$ menu. The $\boxed{\substack{\text{STAT} \\ \text{PLOT}}}$ menu will automatically set the Xlist to L_1 and the Ylist to L_2.

- The logarithmic regression and the power regression may not work, depending on the example. For instance, 0 hours studying produced a 40 on the test, but since the *x*-coordinate is zero these regressions cannot be done.

Dis and Data

The students at West High School have recorded data showing a positive relationship between the number of hours a student studies and the student's grade.

Chart A

Hours	0	0	1	2	2	4	5	6
Grades	40	50	60	65	70	80	90	95

1. Prepare a scatter plot on graph paper.

2. Based on the scatter plot, what relationship exists between the hours studied and the student's grades?

   Using Chart A, enter the data in the $\boxed{\text{STAT}}$ menu and draw a scatter plot. To enter the data, access the $\boxed{\text{STAT}}$ menu, select the DATA submenu, select Edit and enter data points. To draw the scatter plot, access the $\boxed{\text{STAT}}$ menu again and select the DRAW submenu, select Scatter, press $\boxed{\text{ENTER}}$. Adjust the $\boxed{\text{RANGE}}$ as necessary to fit the data.

3. Compare your graph with the graph on the calculator. They should be similar.

4. The *line of best fit* is the line with the same number of data points above it as below it. Find two points that will create a line of best fit and write its equation in slope-intercept form.

Chart B

Student	1	2	3	4	5	6	7	8	9
Shoe Size	8.5	7.5	9.5	11	6.5	8	8.5	7	10
Height	65	69	70	71	62	66	68	65	69
Weight	141	148	164	170	109	150	156	138	165

5. Prepare a scatter plot using height and weight.

6. Describe the relationship that exists between the height and the weight.

7. Describe the change in shoe size as weight increases.

Follow the directions given for Chart A to draw scatter plots for Chart B.

8. Find the equation of the line of best fit comparing height and weight.

9. Find the equation of the line of best fit comparing shoe size and weight.

10. Find the equation of the line of best fit comparing shoe size and height.

11. Collect and prepare data on two related variables, such as

- The number of baskets a basketball player makes versus the number of attempts.

- Age versus the length of a hand

- Grades versus the number of absences

- Length of a hand versus the length of a forearm

    a. Draw a scatter plot and find the equation of the line of best fit.

    b. Test the data using the calculator's statistics features.

        Use the $\boxed{\text{STAT}}$ menu, go to the CALC submenu, select LinReg. To see if the relationship is truly linear, note the values returned by the calculator,

        $a =$

        $b =$

        $r =$

The equation for the line of best fit is $y = bx + a$, where a is the y-intercept and b is the slope. Use r, the *correlation coefficient*, to tell how good the line is. The best lines have r values closest to 1 or −1.

Graphing Power © Dale Seymour Publications

c. Compare the equation generated by linear regression with the line of best fit you found in 11a. Are they the same? How might you arrive at a line of best fit that is closer to the linear regression given by the calculator?

d. Check the other regressions to see if they are better than the linear regression. In the $\boxed{\text{STAT}}$ menu, go to the CALC submenu, try

LnReg (logarithmic regression)	$y = a + b \ln x$ (all $x > 0$)
ExpReg (exponential regression)	$y = ab^x$ (all $y > 0$)
PwrReg (power regression)	$y = ax^b$ (all $x, y > 0$)

e. Choose the best fit equation. $y =$

f. Draw a graph of the best fit equation by going to the $\boxed{\text{DRAW}}$ menu, select DrawF and type in the equation, using the given a and b values from the regression. For example if $a = 0.5$ and $b = -4$, enter DrawF $-4x + .5$ and press $\boxed{\text{ENTER}}$.

g. Organize the information for your data. Include a data chart, a scatter plot, the graph of the best fit equation, a description of the procedure you used, and any conclusions you draw about the relationships in the data from the best fit equation and the r value.

Permutations

Level   Algebra I, Algebra II, Precalculus

Using "Concert Tickets" students will

- Make and test conjectures regarding the calculation of factorials
- Calculate factorials
- Make and test conjectures regarding counting permutations
- Calculate permutations using them $_nP_r$ function on the calculator

Teaching the Lesson

- If your students are not familiar with factorials or need more practice evaluating expressions that contain factorials, ask them to complete the worksheet "Evaluating Factorials."
- This lesson should follow a discussion of the fundamental counting principle and permutations.
- When using the $_nP_r$ function on the calculator, the value for n must be entered before accessing the function.

Solutions for Evaluating Factorials

7. 1

8a. 144
 b. 35
 c. 40
 d. 1512

Solutions for "Concert Tickets"

2. $_nP_r = \dfrac{n!}{(n-r)!}$

5. $_8P_5 = 6720$

7. $_5P_3 = 60$

8. $_{12}P_{10} = 239{,}500{,}800$

11a. $26 \cdot 26 \cdot 26 \cdot 10 \cdot 10 \cdot 10$
  b. $_{26}P_3 \cdot 10 \cdot 10 \cdot 10$
  c. $_{25}P_3 \cdot 10 \cdot 10 \cdot 10$
  d. $_{26}P_3 \cdot 9 \cdot 9 \cdot 9$

Evaulating Factorials

1. For the TI-81, access the [MATH] menu, use the factorial option (!) to complete the chart. For the TI-82, access the [MATH] menu, go to the PBR submenu, select the factorial option (!).

2. Describe the pattern in the chart.

3. How is 3! calculated?

4. How is 6! calculated?

5. How is n! calculated?

n!	Result
1!	
2!	
3!	
4!	
5!	
6!	

6. Test your theory on 7! and 8! Use your calculator to verify that your conjecture was correct.

7!	
8!	

7. 0! is a special number. What do you think 0! is equal to? Use your calculator to check.

8. Find each product or quotient.

   a. $3! \cdot 4!$

   b. $\dfrac{6!}{4!}$

   c. $\dfrac{5! \cdot 2!}{3!}$

   d. $\dfrac{28! \cdot 2!}{26!}$

Concert Tickets

You are going to the concert on Saturday night with four friends. You are really looking forward to the concert, but you have no idea who you will sit next to. How many different ways can you and your four friends sit in the five seats? If there were eight people who wanted to go, but you only had five tickets, how many ways could the five seats be occupied?

The concert problem involves counting the ways to order, or arrange, the elements in a set. Each arrangement is called a *permutation*. The total number of ways to arrange n elements r at a time is described as $_nP_r$

You can find the number of permutations using the $_nPr$ command on your calculator. First enter the value for n, then access the [MATH] menu, go to PRB and select $_nP_r$, enter r, then press [ENTER]. Complete the following chart.

Number of Objects	Number Taken at a Time	n Factorial	(n − r) Factorial	Number of Permutations
n	r	$n!$	$(n-r)!$	$_nP_r$
1	1			
2	2			
2	1			
3	3			
3	2			
3	1			
4	4			
4	3			
4	2			
4	1			
5	5			
5	4			
5	3			
5	2			
5	1			

Graphing Power © Dale Seymour Publications

1. Describe the pattern you see in the last three columns of the chart.

2. Write the expression using n and r and factorials that describe this pattern of total number of permutations of n objects taken r at a time.

3. How many different ways can you and your four friends sit in the five seats?

   a. Is this a *permutation* problem? That is, is it a question about how to arrange elements of a set?

   b How many objects are there?

   c. How many objects we are arranging?

   d. In $_nP_r$ notation, write the expression that counts the number of permutations.

   e. Use your calculator to compute the number of arrangements.

   f. How many people can you choose to sit in seat 1?

   g. How many are left to choose for seat 2?

   h. How many are left to choose for seat 3?

   i. How many are left to choose for seat 4?

   j. How many are left to choose for seat 5?

4. Explain how the $_nP_r$ notation works.

5. If there were eight people who wanted to go, but you only had five tickets, how many ways could the five seats be occupied?

    a. Is this a *permutation* problem? That is, is it a question about how to arrange elements of a set?

    b. How many objects?

    c. How many objects we are arranging?

    d. In $_nP_r$ notation, write the expression that counts the number of permutations.

    e. Use your calculator to compute the number of arrangements.

    f. How many people can you choose to sit in seat 1?

    g. How many are left to choose for seat 2?

    h. How many are left to choose for seat 3?

    i. How many are left to choose for seat 4?

    j. How many are left to choose for seat 5?

6. Explain how the $_nP_r$ notation works.

7. How many different three letter patterns can be formed using the letters A, B, C, D, and E?

8. There are 12 students in a class that seats 10. How many different seating arrangements are possible?

Graphing Power © Dale Seymour Publications

9. Write your own problem using permutations where $n \neq r$.

10. Find a real-life situation where permutations are applied.

11. Find the number of possible license plates, each made up of 3 letters and 3 numbers, for each case.

   a. Any three letters and any three numbers can be used.

   b. No letter may be repeated.

   c. No letter may be repeated, and the letter O can't be used.

   d. No letter may be repeated, and the number 0 can't be used.

Combinations

Level   Algebra I, Algebra II, Precalculus

Using "Committee Time" students will

- Learn to count combinations
- Make and test conjectures regarding combinations
- Calculate factorials
- Calculate combinations using the $_nC_r$ function on the calculator

Teaching the Lesson

- If students haven't completed "Concert Tickets," have them complete the page "Evaluating Factorials," page 39.
- This should follow a brief discussion on the fundamental counting principle, permutations, and combinations.
- When using the $_nC_r$ function on the calculator, the value for n must be entered before accessing the function.

Solutions

2. $\dfrac{n!}{r!(n-r)!}$

13. 2080 connections

14c. 40

Committee Time

You are going to choose a committee of three students from five officers of the student council to serve as the nominating committee for officers for the next school year. How many different nominating committees can be formed?

The committee problem involves counting the ways to choose a set; the order within the set does not matter. A committee of Juan, Paul, and Jennifer is the same as a committee of Paul, Jennifer, and Juan. Each set is called a *combination*. The total number of ways to choose from n elements a set of r elements is described as $_nC_r$.

To use the $_nC_r$ command on your calculator, first enter the value for n, then go to the MATH menu, select the PRB submenu, and select $_nC_r$, then enter the value for r, press ENTER to evaluate this amount. Use the $_nC_r$ function to complete the following chart.

Number of Objects	Number Taken at a Time	n Factorial	r Factorial	$(n-r)$ Factorial	Number of Combinations
n	r	$n!$	$r!$	$(n-r)!$	$_nC_r$
1	1				
2	2				
2	1				
3	3				
3	2				
3	1				
4	4				
4	3				
4	2				
4	1				
5	5				
5	4				
5	3				
5	2				
5	1				

1. Describe the pattern in the chart.

2. Write an expression using n and r and factorials that describes this pattern of total number of combinations of n objects taken r at a time.

3. Is the committee problem a combination problem? That is, is it a question about how to arrange elements from a set without regard to the order?

4. How many objects, elements?

5. How many elements we are choosing?

6. In $_nC_r$ notation, write the expression that counts the number of combinations.

7. Use your calculator to compute the number of arrangements.

8. How many choices are there for the first committee member?

9. How many are left for the second committee member?

10. How many are left for the third committee member?

11. Explain how the $_nC_r$ notation works.

Graphing Power © Dale Seymour Publications

12. How many ways can a four-person committee for the science fair be chosen from the thirteen members of the science club?

    a.   How many choices are there for the first committee member?

    b.   How many are left to choose for the second committee member?

    c.   How many are left to choose for the third committee member?

    d.   How many are left to choose for the fourth committee member?

    e.   How many arrangements of four people comprise the same committee?

13. There are 65 telephones at Jefferson High School. How many two-way connections can be made among the school phones?

14. From the mathematics club officers of four boys and five girls, how many committees of three boys and two girls can be formed?

    a.   How many ways can the boys be chosen? Explain.

    b.   How many ways can the girls be chosen? Explain.

    c.   How many ways can both of these events be chosen together?

15. From a club of 8 boys and 10 girls, how can a nominating committee of 5 be formed in each of the following cases? Explain your reasoning in each case.

    a.   All committee members are boys.

    b.   There are 3 boys and 2 girls on the committee.

    c.   All committee members are girls.

16. In what ways are permutations and combinations the same?

TOPIC 5

Functions

Drawing a Linear Graph Given Two Points

Level   Pre-Algebra, Algebra

Using "Get to the Point!" students will

- Find the slope-intercept form for the equation of a line through two points
- Plot two points using the calculator
- Graph equations using the Y= menu
- Use two other points on the line to verify the equation
- Substitute values for *x* and *y* to verify the equation

Teaching the Lesson

- The directions for plotting points and graphing equations are given on the student page. There is a slight difference in the keystrokes on the TI-82. The commands are found in the DRAW menu on both calculators.
- The DRAW and Y= menus must be cleared before each new problem.
- An understanding of how to calculate slope given two points is helpful.

Solutions

1. $y = 2x + 3$ 2. $y = 3x - 4$

3. $y = -2x + 6$ 4. $y = 3x - 3$

Get to the Point!

	TI-81	TI-82
Suggested Range	$X_{min} = -23.5$	$X_{min} = -23$
	$X_{max} = 24$	$X_{max} = 24$
	$X_{scl} = 1$	$X_{scl} = 1$
	$Y_{min} = -15.5$	$Y_{min} = -15$
	$Y_{max} = 16$	$Y_{max} = 16$
	$Y_{scl} = 1$	$Y_{scl} = 1$
	$X_{res} = 1$	

Follow these steps to plot points and the line that contains those points.

Step 1   Use the $\boxed{\text{DRAW}}$ menu and PT-On command, to plot the two points if using the TI-81. The PT-ON command is in the Points submenu in the $\boxed{\text{DRAW}}$ menu on the TI-82. Plot each point individually and press $\boxed{\text{ENTER}}$. Your points should appear on the coordinate grid.

Step 2   To draw the line that passes through those two points, go to the $\boxed{\text{DRAW}}$ menu, select the DrawF command and type in your expression, and press $\boxed{\text{ENTER}}$. For example, DrawF $2x + 3$ will graph the line $y = 2x + 3$. If your equation is correct, the line will pass through your points. If not, go back to the DrawF command and try again.

Step 3   Use the arrow keys to move the cursor along the line and verify that your points are on the line. Find two other points on the line.

Step 4   Use your equation to verify that the points are on the line by substituting the value for x and finding y.

Step 5   After each problem, be sure to ClrDraw.

Follow steps 1–5 for each pair of points.

1. $(2, 7)$ and $(-3, -3)$

equation $y = $ _____

two other points   (_____,_____)

(_____,_____)

verify

2. (3, 5) and (–2, –10)

     equation   y = _____

     two other points   (____,____)

                                    (____,____)

     verify

3. (4, –2) and (–2, 10)

     equation   y = _____

     two other points   (____,____)

                                    (____,____)

     verify

4. (4, 9) and (–1, –6)

     equation   y = _____

     two other points   (____,____)

                                    (____,____)

     verify

5. Use the two other points that you found in problem 4 to recalculate the equation of the line. How does this compare to the first equation? Will this always be true for any two points on the line? What can you conclude about finding an equation given points on the line?

Graphing Power © Dale Seymour Publications

Graphs of Linear Functions

Level   Pre-Algebra, Algebra I

Using "On Line" the student will

- Use the graphing capabilities to graph linear equations

- Describe how the constants in the equation $y = ax + b$ change the graph of $y = x$

Teaching the Lesson

- Clear all $\boxed{Y=}$ entries prior to graphing each new set of equations.

- The calculator will graph the equations in order. Y_1 will be graphed first, followed by Y_2, etc.

- By deselecting the other equations in the $\boxed{Y=}$ menu, the calculator will graph only one equation at a time. To deselect equations, move the blinking cursor over the = sign and press $\boxed{\text{ENTER}}$. The = is no longer highlighted and will not be graphed on the coordinate grid. To reselect the equation, move the blinking cursor back onto the = sign and press $\boxed{\text{ENTER}}$. The = is again highlighted and will be graphed on the coordinate grid.

Suggested Range

$$
\begin{aligned}
X_{min} &= -10 \\
X_{max} &= 10 \\
X_{scl} &= 1 \\
Y_{min} &= -10 \\
Y_{max} &= 10 \\
Y_{scl} &= 1 \\
X_{res} &= 1
\end{aligned}
$$

On Line

Enter each set of equations into the $\boxed{Y=}$ menu and $\boxed{\text{GRAPH}}$ them one at a time on the same graph grid. Draw a sketch of the lines on the graph grid provided.

1. Graph

   $Y_1 = x$

   $Y_2 = x + 2$

   $Y_3 = x + 6$

   $Y_4 = x - 3$

   a. How are these lines alike?

   b. Where does each line cross the y-axis?

   c. What happens to the graph of $y = x$ when a constant is added? (What does b do in $y = x + b$?)

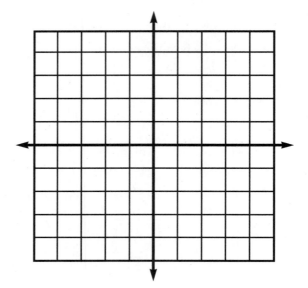

2. Graph

   $Y_1 = x$

   $Y_2 = 2x$

   $Y_3 = 5x$

   $Y_4 = 10x$

   a. How are these lines alike?

   b. How are these lines different?

   c. What happens to the graph of $y = x$ when x is multiplied by a positive number greater than 1? (What does a do in $y = ax$ when $a > 1$?)

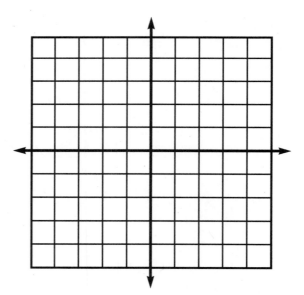

Graphing Power © Dale Seymour Publications

3. Graph

$Y_1 = x$
$Y_2 = 3x$
$Y_3 = -2x$
$Y_4 = -5x$

a. How are these lines alike?

b. How are these lines different?

c. What happens to the graph of $y = x$ when x is multiplied by a negative number less than -1? (What does a do in $y = ax$ when $a < -1$?)

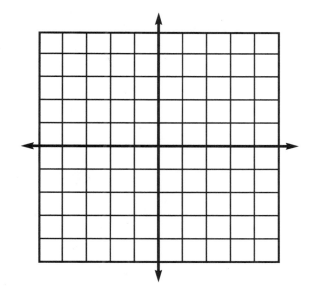

4. Graph

$Y_1 = x$

$Y_2 = \frac{1}{2}x$

$Y_3 = \frac{1}{3}x$

$Y_4 = \frac{1}{5}x$

(Remember to enclose the fractions in parentheses.)

a. What happens to the graph of $y = x$ when x is multiplied by a positive number between 0 and 1? (What does a do in $y = ax$ when $0 < a < 1$?)

b. What will happen to the graph of $y = x$ if x is multiplied by a number between -1 and 0? Edit your functions by placing a negative sign in front of the fraction and graph to test your conjecture.

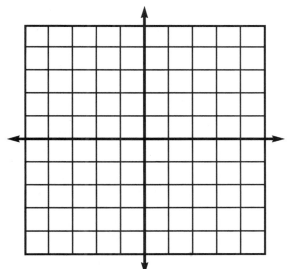

5. Graph
$Y_1 = x$
$Y_2 = 2x$
$Y_3 = 2x + 3$
$Y_4 = -2x - 7$

a. Explain what the constants 2 and –2 do to the graph of $y = x$.

b. Explain what the constants 3 and –7 do to the graph of $y = x$.

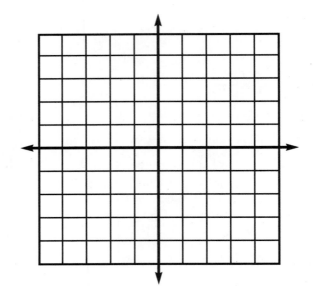

6. Graph
$Y_1 = x$

$Y_2 = \frac{1}{2}x$

$Y_3 = \frac{1}{2}x + 2$

$Y_4 = -\frac{1}{2}x - 4$

a. Explain how the constants $\frac{1}{2}$ and $-\frac{1}{2}$ affect the graph of $y = x$.

b. Explain how the constants 2 and –4 affect the graph of $y = x$.

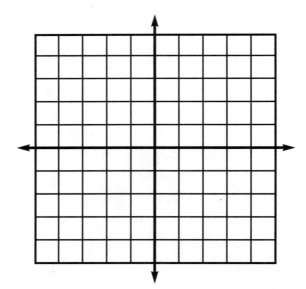

7. For each equation, sketch the graph without the calculator. Then use the graphing calculator to check your graph and modify it if necessary.

a. $y = 3x - 2$

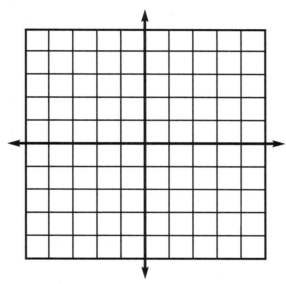

b. $y = -3x + 2$

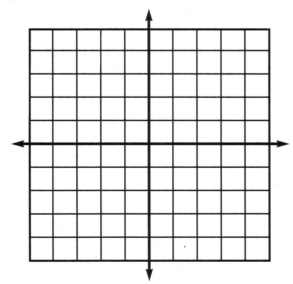

c. $y = \frac{1}{3}x + 4$

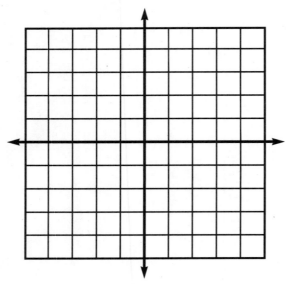

8. Create an equation of your own in the form of $y = ax + b$ and explain how each constant affects the graph of $y = x$.

9. Explain how a and b effect the graph of $y = x$ in the equation $y = ax + b$.

Graphing Power © Dale Seymour Publications

Systems of Equations

Level   Algebra I, Algebra II, Geometry

Using "Systems of Equations" students will

- Use the graphing capabilities to find the solutions of systems of equations
- Write equations in slope-intercept form
- Verify solutions to systems of equations
- Find a system of equations with a given solution

Teaching the Lesson

- Students must be able to change equations into slope-intercept form.

	TI-81	TI-82
Suggested Range	$X_{min} = -9.4$	$X_{min} = -9.4$
	$X_{max} = 9.6$	$X_{max} = 9.6$
	$X_{scl} = 1$	$X_{scl} = 1$
	$Y_{min} = -6.2$	$Y_{min} = -6.2$
	$Y_{max} = 6.4$	$Y_{max} = 6.2$
	$Y_{scl} = 1$	$Y_{scl} = 1$
	$X_{res} = 1$	

Solutions

1. (2, 1)

2. (–2, 2)

3. (0, 5)

4. (2, 2)

6a. yes
  b. yes
  c. no

Sequence of Equations

Rewrite each equation in slope-intercept form ($y = ax + b$). Graph each system of equations. TRACE to find the point of intersection. Verify the solution by substituting the values of x and y into both equations.

1. $x + y = 3$ $x - y = 1$

   Point of intersection (_____,_____)

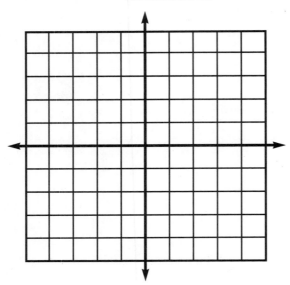

2. $x = -y$ $x + y = 4$

   Point of intersection (_____,_____)

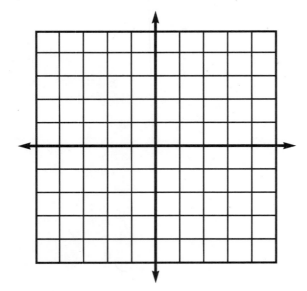

Graphing Power © Dale Seymour Publications

3. $-3x = 5 - y$ $2y = 6x + 10$

Point of intersection (_____,_____)

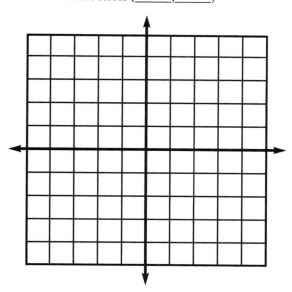

4. $x = \dfrac{1}{2}y + 1$ $x - 2y = -2$

Point of intersection (_____,_____)

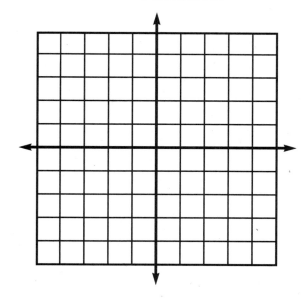

5. How do you know that the point of intersection is the solution to the system of equations?

6. Enter each of the following sets of equations into the $\boxed{\text{Y=}}$ menu and study the graphs. Determine whether the given ordered pair is the solution of the system of equations. (Remember to clear your entries after each set.)

    a. $(4, 2)$ $x - y = 2$ $x + y = 6$

    b. $(-1, 1)$ $x = -1$ $x - y = -2$

    c. $(-1, -3)$ $x = \frac{1}{3}y + 2$ $-2x - y = 1$

7. Find a system of equations with $(2, -4)$ as the solution.

8. Using $5x + 2y = 11$, find another equation such that the solution to the system is $(3, -2)$.

Graphing Power © Dale Seymour Publications

Linear Inequalities

Level   Pre-Algebra, Algebra I

Using "Shades" students will

- Predict solutions of linear inequalities
- Graph linear inequalities
- Test possible solutions of linear inequalities
- Make and prove conjectures regarding linear inequalities
- Write inequalities given a graph

Teaching the Lesson

- Clear the ⎡DRAW⎤ and ⎡Y=⎤ menus before starting each new inequality by selecting the ClrDraw option in the ⎡DRAW⎤ menu.
- When using the shade command, the values in the parentheses will automatically be entered as (minimum value, maximum value).
- Sequence of keystrokes and placement of data and commas are crucial.
- The ⎡Y-VARS⎤ menu allows students to directly enter their equations from the ⎡Y=⎤ menu without retyping them.
- This lesson requires students to write definitions, check their definitions carefully. Student responses will vary.

Solutions

20a. $y < -1$ b. $y < x - 3$

Shades

	TI-81	TI-82
Suggested Range	$X_{min} = -9.4$	$X_{min} = -9.4$
	$X_{max} = 9.6$	$X_{max} = 9.4$
	$X_{scl} = 1$	$X_{scl} = 1$
	$Y_{min} = -6.2$	$Y_{min} = -6.2$
	$Y_{max} = 6.4$	$Y_{max} = 6.2$
	$Y_{scl} = 1$	$Y_{scl} = 1$
	$X_{res} = 1$	

1. Consider the linear inequality $y < x + 2$. Choose three solutions in the form of (x, y) that make the inequality true.

    To verify your solutions on the calculator, enter $y = x + 2$ as Y_1 in the $\boxed{Y=}$ menu, press $\boxed{QUIT}$. Access the $\boxed{DRAW}$ menu and select the Shade command. With the Shade(prompt on the screen, enter the minimum value, a comma, then the maximum value, close the parentheses. Your screen should look similar to this Shade($-6.4,Y_1$). Press $\boxed{ENTER}$. (Y_1 is in the $\boxed{Y\text{-VARS}}$ menu. If using the TI-82, select the Function sub-menu in the $\boxed{Y\text{-VARS}}$ menu.)

2. This is the graph of $y < x + 2$. Where are the three points you choose in question 1? Are they in the shaded or unshaded region?

3. Describe what $y < x + 2$ means.

4. What does the calculator command Shade(Y_{min},Y_1) mean?

5. Use the arrow keys to locate points that are and are not solutions to the inequality $y < x + 2$.

Are	Are Not

Graphing Power © Dale Seymour Publications

6. Verify that these points are or are not solutions to the inequality by substituting for x, comparing to y, and deciding if the $<$ is true or false.

7. Further refine your description of $y < x + 2$.

8. How does $y \leq x + 2$ differ from $y < x + 2$?

9. Consider $y \geq x - 4$. Choose three solutions for the inequality in the form of (x, y) that make the inequality true.

   To verify your solutions, Shade(Y_1,6.4).

10. This is the graph of $y \geq x - 4$. Where are the three points you choose in question 9? Are they in the shaded or unshaded region? Describe what $y \geq x - 4$ means.

11. What does the calculator command Shade(Y_1,Y_{max}) mean?

12. Use the arrow keys to locate points that are and are not solutions to the inequality $y \geq x - 4$.

Are	Are Not

Graphing Power © Dale Seymour Publications

13. Verify that these points are or are not solutions to the inequality by substituting for *x*, comparing to *y*, and deciding if the $\geq$ is true or false.

14. Further refine your description of $y \geq x - 4$.

15. How is $y \geq x - 4$ different from $y > x - 4$?

Use what you know to sketch graphs of the following. Sketch the graph, then graph on the calculator to verify your solution.

16. $y < x + 3$

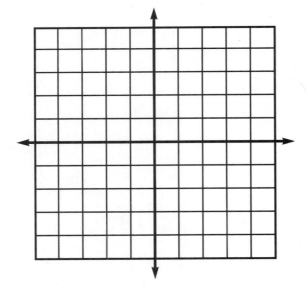

17. $y \geq x + 4$

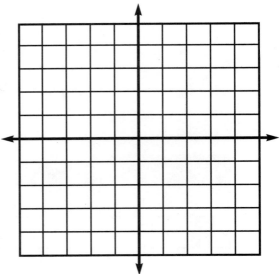

Graphing Power © Dale Seymour Publications

18. $y > 1 - 3x$

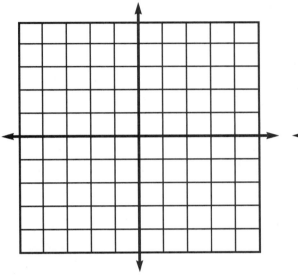

19. $y + 3 < x + 8$

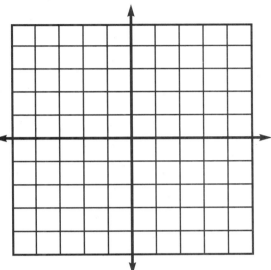

20. Write the inequality whose graph is shown. Verify your inequality on the graphing calculator.

a.

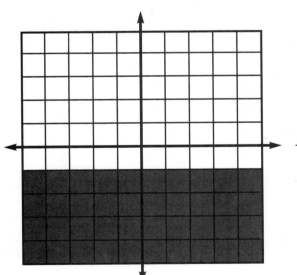

b.

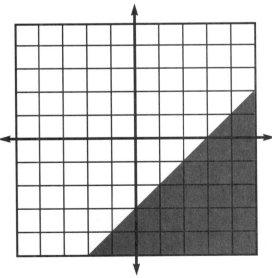

Systems of Linear Inequalities

Level   Pre-Algebra, Algebra I, Algebra II

Using "Shades II" students will

- Predict solutions of systems of linear inequalities

- Graph systems of linear inequalities

- Test possible solutions to systems of linear inequalities

- Make and test conjectures regarding systems of linear inequalities

- Write a system of linear inequalities given a graph

Teaching the Lesson

- Clear the $\boxed{\text{DRAW}}$ and $\boxed{\text{Y=}}$ menus before starting each new inequality by selecting the ClrDraw option in the $\boxed{\text{DRAW}}$ menu.

- When using the shade command, the values in the parentheses will automatically be entered as (minimum value, maximum value).

- Sequence of keystrokes and placement of data and commas are crucial.

- The $\boxed{\text{Y-VARS}}$ menu allows students to directly enter their equations from the $\boxed{\text{Y=}}$ menu without retyping them.

Solutions

11a. $y > x - 3$ b. $y < x + 3$

   $y < 1$ $y > \dfrac{7}{4}x - \dfrac{1}{2}$

Shades II

	TI-81	**TI-82**
Suggested Range	$X_{min} = -9.4$	$X_{min} = -9.4$
	$X_{max} = 9.6$	$X_{max} = 9.4$
	$X_{scl} = 1$	$X_{scl} = 1$
	$Y_{min} = -6.2$	$Y_{min} = -6.2$
	$Y_{max} = 6.4$	$Y_{max} = 6.2$
	$Y_{scl} = 1$	$Y_{scl} = 1$
	$X_{res} = 1$	

Consider this system of linear inequalities. $y \geq 2x + 3$
$y \leq x + 2.$

1. Choose one solution that satisfies both equations.

   To verify your solution, enter $y = 2x + 3$ as Y_1 and $y = x + 2$ as Y_2 in the ⎡Y=⎤ menu. Press ⎡QUIT⎤. Access the ⎡DRAW⎤ menu, select the Shade(command. Enter Shade(y_1,y_2) and press ⎡ENTER⎤. (Y_1 and Y_2 are in the ⎡Y-VARS⎤ menu on the TI-81. If using the TI-82, Y_1 and Y_2 are in Functions submenu in the ⎡Y-VARS⎤ menu.)

2. This is the graph of $y \geq 2x + 3$
   $y \leq x + 2.$

   What does this graph represent? Describe what this system means.

3. Use the arrow keys to locate points that are and are not solutions to the system of inequalities.

Are	Are Not

4. Verify that these points are or are not solutions by substituting for x, comparing to y, and deciding if the point is in the intersection of the inequalities. To be a point in the solution, both inequalities have to be true.

5. Further refine your description of the system
$$y \geq 2x + 3$$
$$y \leq x + 2$$

6. How is the system $\begin{array}{l} y \geq 2x + 3 \\ y \leq x + 2 \end{array}$ different from the system $\begin{array}{l} y > 2x + 3 \\ y < x + 2 \end{array}$?

Use what you know to sketch the graphs of the following systems, then verify on the calculator.

7. $y \leq x + 1$
   $y \geq 2 - x$

8. $y \leq 5x + 3$
   $y - 5 \geq -5x$

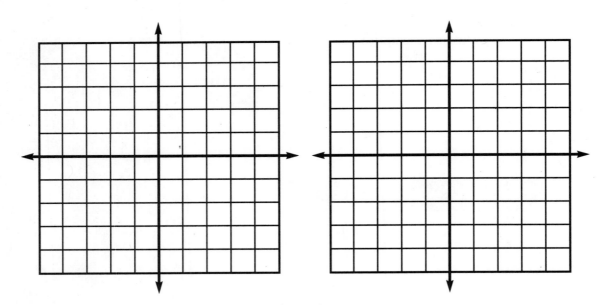

Graphing Power © Dale Seymour Publications

9. $y > 2x - 3$
$y > 2x + 6$

10. $3x - y > -1$
$x - y > -4$

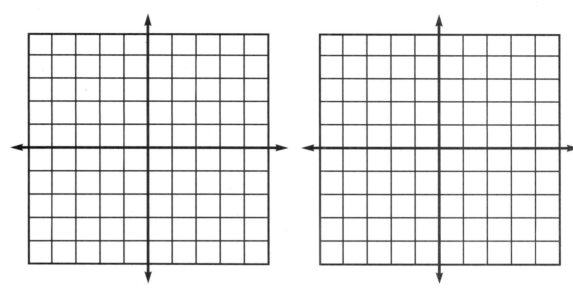

11. Write the system of inequalities whose graph is shown. Verify your system on the graphing calculator.

a.   b.

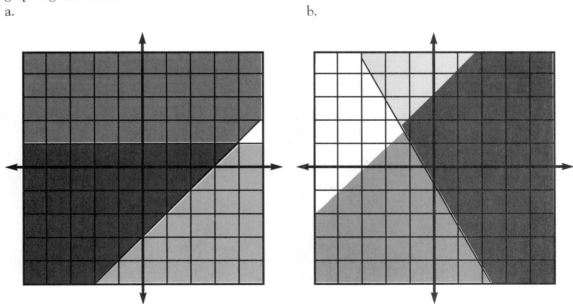

Cramer's Rule

Level Algebra I, Algebra II

Using "Cramer Solves Systems" students will

- Discuss and demonstrate Cramer's Rule
- Demonstrate calculator solutions using Cramer's Rule
- Use Cramer's Rule to solve problems
- Use the $\boxed{\text{MATRX}}$ menu and its capabilities to solve systems

Teaching the Lesson

- The historical information on Cramer's Rule is taken from *Historical Topics for the Mathematics Classroom*, 31st Yearbook of the NCTM. Reston, Va.: National Council of Teachers of Mathematics, 1969.

- The $\boxed{\text{MATRX}}$ menu is very simple to use. Students using the TI-81 may experience some confusion because the data entry forces them to enter down a single column while the actual data from the matrix is read across the rows.

- Remind students that a determinant is not the same as a matrix. Additionally, division by zero is undefined.

Solutions

1. $x = -45$
   $y = -15$

2. $x = 7$
   $y = -5$

3. $x = 12.75$
   $y = 20.5$

4. $L = 73.6$
   $S = 18.4$

5. $J = \$157.50$
   $M = \$52.50$

6. no solutions

Cramer Solves Systems

An 18th century Swiss mathematician, Gabriel Cramer, is credited with discovering a pattern which solves simultaneous equations. However, a Japanese mathematician, Seki Kowa, actually discovered this method in 1683, and, ten years later, it was written in algebraic form by Godfried, Leibniz, and L'Hôpital. Cramer was credited with the discovery when he published this method in 1750, and hence, it is known as Cramer's Rule.

Cramer's Rule uses *determinants* to solve systems of simultaneous equations.

For two equations with two unknown variables, identify the coefficients of the variables, x and y, and put them in determinant form. This gives the "denominator" determinant.

Example: $2x - y = 6$
$-3x + 4y = 3$
becomes $D = \begin{vmatrix} 2 & -1 \\ -3 & 4 \end{vmatrix}$

To solve for x, choose the coefficients for x and replace them with the constant terms and place them in the determinant D. Call this new determinant the X determinant. Similarly, to solve for y, choose the coefficients for y and replace them with the constant terms and place them in the determinant D. This new determinant is called the Y determinant.

$X = \begin{vmatrix} 6 & -1 \\ 3 & 4 \end{vmatrix}$
$Y = \begin{vmatrix} 2 & 6 \\ -3 & 3 \end{vmatrix}$

To solve for x and y, calculate the following determinant ratios.

$x = \dfrac{X}{D} = \dfrac{\begin{vmatrix} 6 & -1 \\ 3 & 4 \end{vmatrix}}{\begin{vmatrix} 2 & -1 \\ -3 & 4 \end{vmatrix}}$
$y = \dfrac{Y}{D} = \dfrac{\begin{vmatrix} 2 & 6 \\ -3 & 3 \end{vmatrix}}{\begin{vmatrix} 2 & -1 \\ -3 & 4 \end{vmatrix}}$

A determinant is not the same as a matrix. However, a determinant can be found from a matrix. That is, $D \neq [A]$, but $D = \det [A]$.

To solve the determinants, use the difference of the cross products:

$D = \begin{vmatrix} 2 & -1 \\ -3 & 4 \end{vmatrix} = (2 \cdot 4) - (-3 \cdot -1) = 5$

$X = \begin{vmatrix} 6 & -1 \\ 3 & 4 \end{vmatrix} = (6 \cdot 4) - (3 \cdot -1) = 27$

$Y = \begin{vmatrix} 2 & 6 \\ -3 & 3 \end{vmatrix} = (2 \cdot 3) - (-3 \cdot 6) = 24$

Therefore,

$$x = \frac{X}{D} = \frac{\begin{vmatrix} 6 & -1 \\ 3 & 4 \end{vmatrix}}{\begin{vmatrix} 2 & -1 \\ -3 & 4 \end{vmatrix}} = \frac{27}{5} \qquad y = \frac{Y}{D} = \frac{\begin{vmatrix} 2 & 6 \\ -3 & 3 \end{vmatrix}}{\begin{vmatrix} 2 & -1 \\ -3 & 4 \end{vmatrix}} = \frac{24}{5}$$

These are the solutions to the system of equations. Verify these solutions by substituting the values for x and y into both equations.

The graphing calculator and Cramer's rule can be used to solve systems of equations.

- Access the [MATRX] menu, go to the EDIT submenu, press [ENTER]. Select [A] and make [A] a 2×2 matrix.

- Use the example above and fill in [A] using the coefficients for D. Enter the values row by row.

- [QUIT] returns you to the Home Screen where you can see [A], which contains your D values. To view [A], press [A] [ENTER]. If using the TI-82, [A] can be viewed by selecting Name in the [MATRX] menu.

- Follow the same procedure to enter your x coefficients into [B] and y coefficients into [C] in a similar fashion.

- To solve for x and y, we need to find the value of the determinants.

  To find D, access the [MATRX] menu, select det, enter [A], press [ENTER]. If using the TI-82, access the [MATRX] menu, go to the MATH submenu, and select det, press [MATRX], select [A], and press [ENTER]. Your answer should be 5.
  (If D = 0, then there are no solutions since division by zero is impossible.)

  To find X and Y, find the determinants of [B] and [C]. You should get X = 27 and Y = 24.

- To find x and y, divide the determinants.

  You can find and divide the determinates on the calculator in one step:

  det [B] [÷] det [A] [ENTER]. As a verification, you should have $x = 5.4$ and $y = 4.8$.

Graphing Power © Dale Seymour Publications

Solve these systems of equations using the MATRX functions.

1. $x - 3y = 0$
$x + y = -60$

$x =$ _____

$y =$ _____

2. $5x + y = 30$
$3x - 4y = 41$

$x =$ _____

$y =$ _____

3. $2x + y = 46$
$8x - 4y = 20$

$x =$ _____

$y =$ _____

Solve

4. Lisa weighs four times as much as her baby sister. Together they weigh 92 pounds. How much does each girl weigh?

5. Joe earned three times as much as Marty. Together they earn $210. How much did each boy make?

6. Tony buys 39 pizzas and 21 gallons of soda for $396.00. Ann gets 52 pizzas and 28 gallons of soda for $518.00. Assuming the same costs for the items, is this possible? Explain.

Motion Problems

Level   Algebra I, Algebra II

Using "Two Trains" students will

- Use the parametric setting to graph the motion of a problem
- Use the [TRACE] feature to investigate the results of the problem

Teaching the Lesson

- It is suggested that students work in groups to find the following expressions and range values. Some groups may need help determining the equations.

Train 1		Train 2
$X_{1T} = 50T$	{distance}	$X_{2T} = 55(T-1)$
$Y_{1T} = 1$	{track}	$Y_{2T} = 2$

Range

$0 \leq T \leq 15$ {15 hours to travel from 750 miles at 50 mph}

$0 \leq X \leq 750$ {Beaumont is 750 miles away}

$0 \leq Y \leq 3$ {we need only 2 tracks}

- In the [MODE] menu, select PARAMetric and SIMULtaneously.

Solutions

7. Train 2 arrives in Beaumont in 14.65 hours.
   Train 1 arrives in 15 hours.

8. Change to a value less than 0.1.

Two Trains

Two trains travel from El Paso to Beaumont (750 miles). Train 1, traveling 50 mph, leaves an hour before Train 2, which maintains an average speed of 55 mph. Which train arrives at Beaumont first?

1. Write the expression for the distance traveled by Train 1 at any given time T. In the $\boxed{Y=}$ menu, enter this equation for X_{1T}. Enter 1 for Y_{1T} as this train will travel on track 1.

2. Write the expression for the distance traveled by Train 2 at any given time T. In the $\boxed{Y=}$ menu, enter this equation for X_{2T}. Enter 2 for Y_{2T} as this train will travel on track 2.

Determine the range necessary to observe the trains.

3. How long does it take the slowest train to reach Beaumont?

    Therefore, the range of T (time) should be from _____ to _____.

4. The range of X, the distance traveled, should be from _____ to _____.

5. The tracks are represented Y. Why are we using two separate tracks?

6. Enter this information into the $\boxed{\text{RANGE}}$ settings with a T_{step} of 0.1.

7. Press $\boxed{\text{GRAPH}}$.

    Use $\boxed{\text{TRACE}}$ to determine whether or not Train 2 catches up to Train 1 before reaching Beaumont. If so, when? Which gets there first?

8. How could the T_{step} be modified to slow down the trains?

9. Alter the equations so that the trains approach each other from opposite directions. What modifications could be made to make the trains crash?

10. If Train 2 leaves two hours after Train 1, will Train 2 catch Train 1 before it gets to Beaumont? If so, when?

11. Write a problem of your own similar to the train problem, and demonstrate the problem graphically on your calculator.

Graphing Power © Dale Seymour Publications

Inverse Linear Functions

Level   Algebra I, Algebra II, Precalculus

Using "On the Line" students will

- Use the graphing capabilities of the TI-81 and TI-82
- Use ⟨TRACE⟩ to find ordered pairs
- Given a linear function, find its inverse by solving for x
- Compare linear functions and their inverses
- Make and test conjectures based on the calculator displays

Teaching the Lesson

- Students will use the menus
  ⟨Y=⟩, ⟨ZOOM⟩, ⟨GRAPH⟩, and ⟨TRACE⟩.
- Students must be able to solve an equation for both x and y.

Solutions

8. The x and y values are the same, simply reversed.

10. The two lines are reflections through the line $y = x$. Their slopes are reciprocals of each other. The x-intercept of one is the y-intercept of the other.

15. The ordered pairs are inverses of each other. If one line contains the point $(4, 3)$ the other line will contain the point $(3, -4)$.

On the Line

	TI-81	**TI-82**
Suggested Range	$X_{min} = -2$	$X_{min} = -2$
	$X_{max} = 7.5$	$X_{max} = 7.4$
	$X_{scl} = 1$	$X_{scl} = 1$
	$Y_{min} = -2$	$Y_{min} = -2$
	$Y_{max} = 4.3$	$Y_{max} = 4.2$
	$Y_{scl} = 1$	$Y_{scl} = 1$
	$X_{res} = 1$	

Graph the following using the $\boxed{Y=}$ and $\boxed{GRAPH}$ menus on the calculator. Record the graphs on the coordinate grid provided.

1. Enter the equation $Y_1 = x$ and graph.

2. Use the $\boxed{TRACE}$ function to list five ordered pairs below.

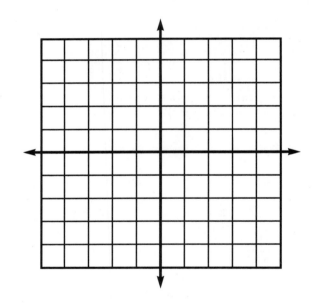

x	y

3. What do you notice about this set of ordered pairs?

4. Graph $Y_2 = 2x + 2$.

5. Solve for x: $\dfrac{y-2}{2} = x$

Let $y = x$ so you have $y = \dfrac{x-2}{2}$ or $y = \dfrac{x}{2} - 1$. Enter and graph this equation as $Y_3 = \dfrac{x}{2} - 1$

Graphing Power © Dale Seymour Publications

6. $\boxed{\text{GRAPH}}$ all three equations. Write a paragraph describing the calculator display.

7. $\boxed{\text{TRACE}}$ to find the ordered pairs listed for each function.

$$y = x \qquad\qquad y = 2x + 2 \qquad\qquad y = \frac{1}{2}x - 1$$

x	y
0	
1	
2	
3	
4	
5	

x	y
0	
1	
2	
3	
4	
5	

x	y
2	
4	
6	
8	
10	
12	

8. What do you notice about the ordered pairs for $y = 2x + 2$ and $y = \frac{1}{2}x - 1$?

9. The equations $y = 2x + 2$ and $y = \frac{1}{2}x - 1$ are inverses. Use the calculator display and the chart above to describe what you notice about these two equations.

10. Why are they are called inverses of each other?

11. Solve $y = \frac{1}{2}x - 1$ for x and compare the results to $y = 2x + 2$. What do you notice?

12. Why are they called inverses of each other?

13. $\boxed{\text{GRAPH}}$ these equations $Y_1 = x$
                                   $Y_2 = 3x - 2$
                                   $Y_3 = $ the inverse of Y_2
                                   (You need to solve for x, then replace y for x.)

    a.   What do you notice about the graphs of the equations?

    b.   What is true about the ordered pairs in Y_2 compared to Y_3? (You can use $\boxed{\text{TRACE}}$ and record the ordered pairs in a chart.)

14. Enter your own set of equations into the $\boxed{\text{Y=}}$ menu and graph.

   Let $Y_1 = x$, $Y_2 = $ your equation, and $Y_3 = $ its inverse.

    a.   What do you notice about the graphs of Y_2 and Y_3?

    b.   What do you notice about the ordered pairs for Y_2 and Y_3? (You can use $\boxed{\text{TRACE}}$.)

    c.   What is the slope of Y_2? What is the y-intercept of Y_2?

    d.   What is the slope of Y_3? What is the x-intercept of Y_3?

    e.   What do you notice about the slopes and intercepts?

15. Given a linear equation describe how to find its inverse. What is true regarding the ordered pairs, slopes, and intercepts?

16. Inverses are said to be *reflections* of each other with the line $y = x$ as the mirror. In a brief paragraph, explain why this is true.

17. Sometimes we test for functions with the *vertical line test*. The vertical line test says that if no vertical line intersects the graph of a given equation in more than one point, then the graph is the graph of a function.

   Do all linear functions have inverses that are functions? Find a nonlinear function that has an inverse that is not a function. (Hint, try $y = x^2$, $y = x^3$, $y = x^4$, etc.)

Graphing Power © Dale Seymour Publications

Solving Quadratics by Graphing

Level   Algebra I, Algebra II

Using "Solving with Graphs" students will

- Enter quadratic equations and discover the meaning of the solutions

- Find the vertex of a quadratic function

- Find the axis of symmetry for a quadratic function

- Write the quadratic equation given a graph

- Use the quadratic formula to verify solutions

Teaching the Lesson

- This lesson assumes students can solve quadratic equations by factoring.

- Remind students to enclose fractional coefficients in parentheses.

- When using the ZOOM function, the TI-82 may convert to scientific notation when giving values for x and y.

- The RANGE may need to be reset after using the ZOOM menu.

Solutions

3. The solutions are the same.

4. They are the same.

9a. $x = 2 \pm \sqrt{10}$

12a. no real solutions
  b. one real solution

Solving with Graphs

Graphing Power © Dale Seymour Publications

	TI-81	TI-82
Suggested Range	$X_{min} = -9.4$	$X_{min} = -9.4$
	$X_{max} = 9.6$	$X_{max} = 9.4$
	$X_{scl} = 1$	$X_{scl} = 1$
	$Y_{min} = -6.2$	$Y_{min} = -6.2$
	$Y_{max} = 6.4$	$Y_{max} = 6.2$
	$Y_{scl} = 1$	$Y_{scl} = 1$
	$X_{res} = 1$	

1. Enter the following equation in the $\boxed{Y=}$ menu: $Y_1 = x^2 - 4$ and $\boxed{GRAPH}$.

   a. Use $\boxed{TRACE}$ to find the x-intercepts $x_1 = $ _____ $x_2 = $ _____

   b. The vertex of the graph is point _____

   c. The equation of the axis of symmetry is _____

2. Use factoring to solve the equation $x^2 - 4 = 0$.

   a. The factors are (_____) and (_____).

   b. The solutions are $x_1 = $ _____ $x_2 = $ _____

3. How do the solutions you found by factoring and the x-intercepts you found on the calculator compare?

4. What can you conclude about the x-intercepts and the solutions?

5. Is there another method to solve a quadratic equation?

6. Enter $Y_1 = -x^2 - 2x + 3$ in the $\boxed{Y=}$ menu and $\boxed{GRAPH}$.

   a. Factor $y = -x^2 - 2x + 3$. The factors are (_____) and (_____).

   b. $\boxed{TRACE}$ to find the x-intercepts $x_1 = $ _____ $x_2 = $ _____

   c. The vertex of the graph is point _____

   d. The equation of the axis of symmetry is _____

7. Use the quadratic equation to solve the equation.

$$x = \frac{-b \pm \sqrt{b^2 - 4ac}}{2a}$$

The solutions are $x_1 =$ _____ $x_2 =$ _____

8. How do your x-intercepts and the solutions compare? Why does this happen?

9. Enter $Y_1 = \frac{1}{2}x^2 - 2x - 3$ in the $\boxed{Y=}$ menu and $\boxed{GRAPH}$. Don't forget to put the fractional coefficient in parentheses.

  a. Use $\boxed{TRACE}$ and $\boxed{ZOOM}$ to find the x-intercepts as accurately as possible.

  $x_1 =$ _____ $x_2 =$ _____

  b. What did you do to get the best value for your x-intercepts?

  c. Show that your x-intercepts and the solution to the equation are the same.

10. Examine the graph.

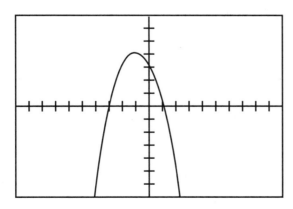

  a. What are the solutions to the equation represented by this graph?

  $x_1 =$ _____ $x_2 =$ _____

b. Use this information to write the binomial equation represented by the graph.

(_____)(_____) = 0

c. Use this information to write the equation represented by the graph.

d. Enter your equation into the $\boxed{Y=}$ menu and $\boxed{GRAPH}$. Does it look like the one pictured? If so, your equation is probably correct. If not, adjust your equation until it matches the graph above.

11. Use the calculator to discover other quadratic equations that have the same solutions as the equation in question 10. (There are many!)

a. How can two different graphs have the same x-intercepts?

b. Can two different graphs have the same equation? Why or why not?

12. Use the calculator to discover the solutions to the following.

a. $x^2 + 2x + 3 = 0$

b. $x^2 - 4x + 4 = 0$

c. Prepare an oral presentation explaining the difference between these equations and the ones used in problems 1–11. Explain what these graphs and their solutions represent.

Graphing Power © Dale Seymour Publications

Graphs of Quadratic Functions

Level   Algebra I, Algebra II, Precalculus

Using "Graphing the Quadratic" the student will

- Use the graphing capabilities to graph quadratic equations

- Observe the differences and similarities among the quadratic graphs

- Describe in writing, how the constants in the equation
  $y = ax^2 + bx + c$ affect the graph of $y = x^2$.

Teaching the Lesson

- Remind students to place fractional coefficients in parentheses.

- The calculator will graph the equations in the order in which they are entered into the $\boxed{\text{Y=}}$ menu.

- To graph Y_1, Y_2, and Y_3 separately. Deselect the equations you do not want graphed by going to the $\boxed{\text{Y=}}$ menu and moving the cursor on top of the darkened = sign and press $\boxed{\text{ENTER}}$. The darkened square is replaced with a regular = sign and the equation will not graph. To re-select the equation, repeat the above instructions. The equation will once again be graphed.

- Conclusions should be drawn and verified as a class before going on to question 4. Inconclusive or incorrect assumptions will make it difficult for the students to be successful on the rest of the assignment.

- $\boxed{\text{RESET}}$ the range to the default settings.

Graphing the Quadratic

Enter each set of equations into the $\boxed{Y=}$ menu and $\boxed{GRAPH}$ them one at a time. Draw a sketch of the equation on the grid provided to answer the following questions.

1. $Y_1 = x^2$
$Y_2 = 3x^2$
$Y_3 = 10x^2$
$Y_4 = \dfrac{1}{2}x^2$

   a. What do all these graphs have in common?

   b. What effect does $a > 1$ have on the graph of $y = ax^2$?

   c. What effect does $0 < a < 1$ have on the graph of $y = ax^2$?

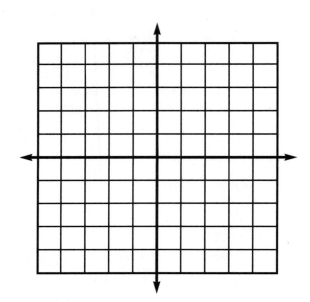

2. $Y_1 = x^2$
$Y_2 = -3x^2$
$Y_3 = -10x^2$
$Y_4 = -\dfrac{1}{2}x^2$

   a. What effect does $a < 0$ have on the graph of $y = ax^2$?

   b. In general, explain what effect a has on the graph of $y = ax^2$.

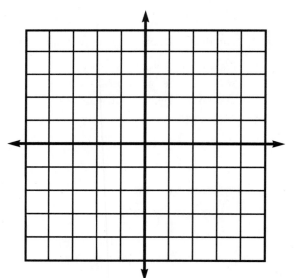

Graphing Power © Dale Seymour Publications

3. $Y_1 = x^2 + 3$
$Y_2 = x^2 - 2$
$Y_3 = x^2 + 6$
$Y_4 = x^2 - 5$

a. What effect does c have on the graph of $y = ax^2 + c$?

b. How is the graph of $ax^2 + c$ affected if $c < 0$? If $c > 0$?

c. If $x = 0$, what is the value of y? What do we call this number?

d. What can you conclude about the significance of c in the general equation?

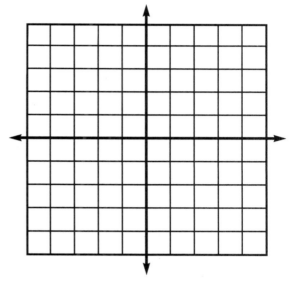

4. $Y_1 = x^2 - 5x$
$Y_2 = x^2 - 2x$
$Y_3 = x^2 + 5x$
$Y_4 = x^2 + 2x$

a. In the equation $y = ax^2 + bx + c$, what effect does the constant b have on the graph?

b. Compare the graphs of $y = x^2 - 2x$ and $y = x^2 + 2x$. Are they congruent? Explain.

c. How is the graph affected if $b > 0$? If $b < 0$?

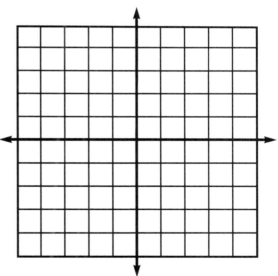

5. Since the value of c in the equation of $y = ax^2 + bx + c$ affected the vertical shift along the axis of symmetry, the equation $y = ax^2 + bx$ can be used to find the axis of symmetry where $x = -\dfrac{b}{2a}$. Show how this can be done using $0 = ax^2 + bx$ and the midpoint formula.

6. Explain, in general, what effect each constant has on the equation $y = ax^2 + bx + c$ as compared to $y = x^2$.

Quadratic Systems

Level   Algebra II, Precalculus

Using "Quadratic System Solutions" students will

- Graph quadratic systems

- Determine the number of solutions from the graph

- Estimate the solutions using the $\boxed{\text{TRACE}}$ function

- Generalize the number of possible solutions and sketch graphs representing the possibilities

Teaching the Lesson

- To graph a circle, the equation for each half of the circle must be entered into the $\boxed{\text{Y=}}$ menu separately. The same is true for the hyperbola. The most efficient way to do so is to enter one equation into Y_1 and to enter the other half of the quadratic into Y_2 by doing the following; access the $\boxed{\text{Y=}}$ menu, press $\boxed{(-)}$, press $\boxed{\text{Y-VARS}}$, and select Y_1, so that $Y_2 = -Y_1$. If using the TI-82, Y_1 is in the function submenu of the $\boxed{\text{Y-VARS}}$ menu.

Solutions

1a.   line, parabola
 b.   2 solutions
 c.   (1, 4) and (-4, 4)

2a.   parabola, circle
 b.   4 solutions
 c.   Answers will vary due to tracing.

3a.   ellipse, hyperbola
 b.   4 solutions

Quadratic System Solutions

Suggested range for TI-81, use $-9.4 \leq x \leq 9.6$ and $-6.3 \leq y \leq 6.3$.
Suggested range for TI-82, use $-9.4 \leq x \leq 9.4$ and $-6.2 \leq y \leq 6.2$.

1. Graph the linear-quadratic system $\quad y = -x + 5$
$$y = x^2 + 2x + 1$$

   a. What is the shape of each graph?

   b. How many solutions are there to this system?

   c. Use the $\boxed{\text{TRACE}}$ function to list the solution pairs for this system.

   d. Are there other possible solutions for a system of linearquadratic equations? If yes,

   draw sketches illustrating each possibility.

2. Graph this system of equations. $\quad y = 3x^2 - 5x - 2$
$$x^2 + y^2 = 9 \quad \text{(Solve for } y \text{, then use each half circle as}$$
   a function in the $\boxed{\text{Y=}}$ menu.)

   a. What is the shape of each graph?

   b. How many solutions are there to this system?

   c. Use the $\boxed{\text{TRACE}}$ function to list the solution pairs for this system.

   d. Are there other possible solutions for a system of this type? If yes, draw sketches illustrating each possibility.

3. Solve for y, then graph this system of equations. $\quad 2x^2 + 3y^2 = 10$
$$-4x^2 + 7y^2 = 6$$

   a. What is the shape of each graph?

   b. How many solutions are there to this system?

   c. What are all possible solutions for a system of this type? Draw sketches illustrating each possibility.

4. Choose two quadratic shapes and sketch the possible ways that they can intersect. Indicate the number of possible solutions for each pair. You may not pick a combination found in questions 1, 2, or 3.

5. Explain what type of graphic solutions can occur when graphing quadratic systems.

Augmented Matrices

Level   Algebra II, Trigonometry, Precalculus

Using "Solve That System" students will

- Create an augmented matrix
- Perform row operations on the augmented matrix
- Record solutions
- Enter the coefficient matrix and the answer matrix into the calculator
- Multiply the inverse of the coefficient matrix and the answer matrix, and record solutions

Teaching the Lesson

- Prior to showing students how to use the calculator to solve these systems, students should have proved if $[C][V] = [A]$, then $[V] = [C]^{-1}[A]$.
- Use the $\boxed{x^{-1}}$ key for $[C]^{-1}$.
- Problem 1 is designed to be done without the calculator.
- For more practice, assign problems from your textbook.
- The calculator can easily solve up to six equations at a time.

Solutions

3. $x = 3$
    $y = -2$
    $z = 1$

9. $x_1 = 5$
    $x_2 = 8$
    $x_3 = 7$
    $x_4 = 4$

10. $x_1 = 1$
    $x_2 = 2$
    $x_3 = -1$
    $x_4 = 3$

11. no solutions

12. $x = 1$
    $y = 0$
    $z = 0$

13. no solutions

Solve that System

1. Use these equations

$$2x + 3y - z = -1$$
$$-x + 5y + 3z = -10$$
$$3x - y - 6z = 5$$

to create a 3×4 augmented matrix

2. Perform row operations to reduce the matrix to

$$\begin{bmatrix} 1 & 0 & 0 & | & \text{value of } x \\ 0 & 1 & 0 & | & \text{value of } y \\ 0 & 0 & 1 & | & \text{value of } z \end{bmatrix}$$

3. Record the results. $x =$ $y =$ $z =$

Augmented matrices can be written 2 ways

$$\begin{bmatrix} c_{11} & c_{12} & c_{13} & | & a_1 \\ c_{21} & c_{22} & c_{23} & | & a_2 \\ c_{31} & c_{32} & c_{33} & | & a_3 \end{bmatrix}$$

or

$$\begin{bmatrix} c_{11} & c_{12} & c_{13} & | & a_1 \\ c_{21} & c_{22} & c_{23} & | & a_2 \\ c_{31} & c_{32} & c_{33} & | & a_3 \end{bmatrix} \begin{bmatrix} x \\ y \\ z \end{bmatrix} = \begin{bmatrix} a_1 \\ a_2 \\ a_3 \end{bmatrix}$$

Call the coefficient matrix C, the variable matrix V and the answer matrix A. Thus we rewrite the equation as

$$CV = A$$

We want to solve for V. That is $V = \begin{bmatrix} x \\ y \\ z \end{bmatrix}$

By the rules of matrix algebra,

$$CV = A$$
$$C^{-1}CV = C^{-1}A$$
$$\text{but } C^{-1}C = I \text{ and } IV = V, \text{ so}$$
$$V = C^{-1}A$$

We can use this identity and the functions of the graphing calculator to solve systems of equations.

4. In the $\boxed{\text{MATRX}}$ menu, go to EDIT and select [C] press $\boxed{\text{ENTER}}$. Make [C] a 3×3 matrix. Enter the numbers from the left portion of the augmented matrix into matrix [C].

5. Go back to the $\boxed{\text{MATRX}}$ and EDIT matrix [A]. Make [A] a 3×1 matrix. Enter the numbers from the right portion of the augmented matrix into matrix [A].

6. $\boxed{\text{QUIT}}$ to return to the Home Screen. Calculate $[C]^{-1} * [A]$ to solve the system. Enter $[C]^{-1} * [A]$. Press $\boxed{\text{ENTER}}$ to calculate the matrix product. Use the $\boxed{x^{-1}}$ for $^{-1}$. If using the TI-82, [C] is in the $\boxed{\text{MATRX}}$ menu in the Name submenu.

7. Record the results, and compare them to the results from question 3.

$x =$ $\qquad$ $y =$ $\qquad$ $z =$

8. Which method was easier? Explain.

Solve the following systems using the procedures in questions 1–6. For systems with more than three variables, it is customary to label the variables $x_1, x_2, x_3, \ldots, x_n$.

9. $\begin{aligned} 9x_1 + 5x_2 + x_3 + x_4 &= 96 \\ 5x_1 + 3x_2 + 7x_3 + 4x_4 &= 114 \\ 5x_1 + 4x_2 + 7x_3 + 3x_4 &= 118 \\ 8x_1 + 7x_2 + 4x_3 + x_4 &= 128 \end{aligned}$

Graphing Power © Dale Seymour Publications

10. $2x_4 - 3x_3 + 2x_1 + x_2 = 13$
$x_4 = 3x_2 - x_1 - 2$
$2x_1 + x_4 - x_2 - x_3 = 4$
$3x_2 = 2x_4$

Questions 11–13 are a examples of special cases. Explain the significance of the solutions, or lack of a solution.

11. Solve this system.

$x + 2y + 3z = 3$
$4x + 5y + 6z = 3$
$7x + 8y + 9z = 3$

12. Solve this system.

$4x + 3y + 5z = 4$
$7x + 8y + 5z = 7$
$3x + 5y + 9z = 3$

13. Solve this system.

$3x + 6y - 12z = 39$
$x + 2y - 4z = 13$
$2x - 4y + z = -8$

Cubic and Quartic Functions

Level   Algebra II

Using "The Shape of It All!" students will

- Use the graphing capabilities to explore cubic and quartic functions
- Determine the effect *a* has on cubic and quartic functions

Teaching the Lesson

- When graphing equations with fractional coefficients, remind students to use parentheses.
- To extend the lesson, ask the students to explore,

$$y = ax^3 + bx^2 + cx + d \text{ and}$$

$$y = ax^4 + bx^3 + cx^2 + dx + e$$

for varying values of *a*, *b*, *c*, *d*, and *e*. As the value of *x* increases or decreases, what happens to the value of *y*?

The Shape of It All!

Enter each of the following equations into the $\boxed{Y=}$ menu and graph one at a time on the same grid. Record the graph of each on the grid provided.

	TI-81	TI-82
Suggested Range	$X_{min} = -4.7$	$X_{min} = -4.7$
	$X_{max} = 4.8$	$X_{max} = 4.7$
	$X_{scl} = 1$	$X_{scl} = 1$
	$Y_{min} = -2$	$Y_{min} = -2$
	$Y_{max} = 3.3$	$Y_{max} = 3.2$
	$Y_{scl} = 1$	$Y_{scl} = 1$

1. To analyze the graph of $y = ax^3$ where $a > 0$, $\boxed{\text{GRAPH}}$ the following.

   $Y_1 = x^{39}$
   $Y_2 = 4x^3$
   $Y_3 = 10x^3$
   $Y_4 = \frac{1}{4}x^3$

   a. What do these graphs have in common?

   b. As the value of x increases, what happens to the value of y?

   c. As the value of x decreases, what happens to the value of y?

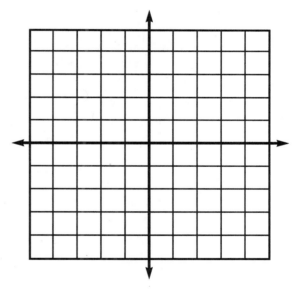

2. To analyze the graph of $y = ax^3$ where $a < 0$, $\boxed{\text{GRAPH}}$ the following.

   $Y_1 = -x^3$
   $Y_2 = -4x^3$
   $Y_3 = -10x^3$
   $Y_4 = -\frac{1}{4}x^3$

   a. What do these graphs have in common?

   b. As the value of x increases, what happens to the value of y?

   c. As the value of x decreases, what happens to the value of y?

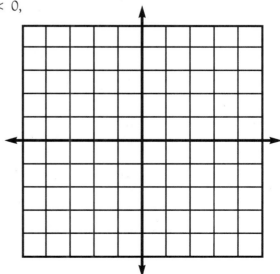

3. To analyze the graph of $y = ax^4$ where $a > 0$, GRAPH the following.

$Y_1 = x^4$

$Y_2 = 5x^4$

$Y_3 = \dfrac{1}{4}x^4$

$Y_4 = \dfrac{1}{10}x^4$

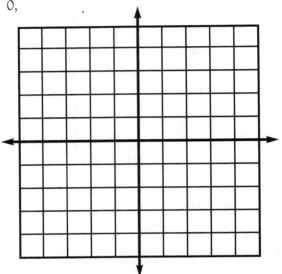

   a. What do these graphs have in common?

   b. As the value of x increases, what happens to the value of y?

   c. As the value of x decreases, what happens to the value of y?

4. To analyze the graph of $y = ax^4$ where $a < 0$, GRAPH the following.

$Y_1 = -x^4$

$Y_2 = -5x^4$

$Y_3 = -\dfrac{1}{4}x^4$

$Y_4 = -\dfrac{1}{10}x^4$

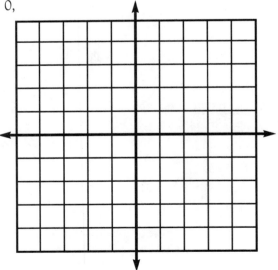

   a. What do these graphs have in common?

   b. As the value of x increases, what happens to the value of y?

   c. As the value of x decreases, what happens to the value of y?

Graphing Power © Dale Seymour Publications

5. Sketch each function. Use the calculator to check your sketch.

$y = -2x^4$

$y = \dfrac{1}{5}x^3$

$y = -3x^3$

$y = \dfrac{3}{2}x^4$

6. Explain why the graphs of $y = ax^3$ and $y = ax^4$ are different. (Explore the effect of positive versus negative values of x in each equation.)

Graphs of Rational Functions

Level   Algebra II, Precalculus

Using "Rational Functions" the student will

- Use the graphing capabilities to explore rational functions
- Observe the differences and similarities among graphs of $y = \dfrac{a}{x + b} + c$ when the constants are changed
- Describe how the constants affect the graph

Teaching the Lesson

- To graph one equation at a time, deselect equations in the $\boxed{\text{Y=}}$ menu by using the cursor arrows to move the cursor on the = sign. Press $\boxed{\text{ENTER}}$. The = sign is no longer highlighted and that equation will not be graphed. To reselect the equation, follow the same procedure. The = sign is now highlighted and that equation will be graphed.

Rational Functions

Enter each set of equations into the $\boxed{Y=}$ menu of the calculator and graph them one at a time on the same graph grid. Sketch each set of equations. Use the general form of $y = \dfrac{a}{x + b} + c$ to answer the questions.

1. Graph
$$Y_1 = \frac{1}{x}$$
$$Y_2 = \frac{1}{x - 1}$$
$$Y_3 = \frac{1}{x - 2}$$

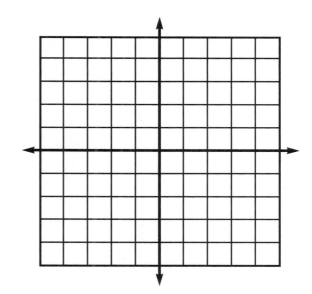

a. In each of the functions, what is the value of a? of b? of c?

b. Which of these values varied?

c. Describe the effect on the graph when this value changed.

2. Graph
$$Y_1 = \frac{1}{x}$$
$$Y_2 = \frac{1}{x + 1}$$
$$Y_3 = \frac{1}{x + 2}$$

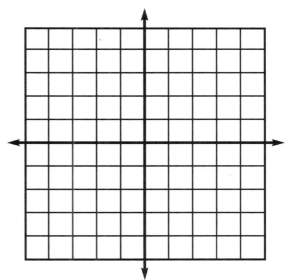

a. In each of the functions, what is the value of a? of b? of c?

b. Which of these values varied?

c. Describe the effect on the graph when this value changed.

3. Graph $\qquad Y_1 = \dfrac{2}{x-1}$

$\qquad\qquad Y_2 = \dfrac{-2}{x-1}$

$\qquad\qquad Y_3 = \dfrac{3}{x-1}$

$\qquad\qquad Y_4 = \dfrac{-3}{x-1}$

a. In each of the functions, what is the value of *a*? of *b*? of *c*?

b. Which of these values varied?

c. Describe the effect on the graph when this value changed.

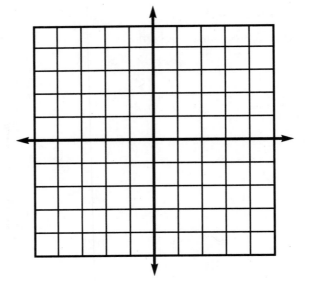

4. Graph $\qquad Y_1 = \dfrac{1}{x+2}$

$\qquad\qquad Y_2 = \dfrac{2}{x+2}$

$\qquad\qquad Y_3 = \dfrac{0.3}{x+2}$

a. In each of the functions, what is the value of *a*? of *b*? of *c*?

b. Which of these values varied?

c. Describe the effect on the graph when this value changed.

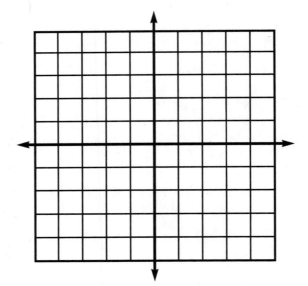

Graphing Power © Dale Seymour Publications

5. Graph

$$Y_1 = \frac{2}{x-1} + 1$$

$$Y_2 = \frac{-2}{x-1} - 1$$

$$Y_3 = \frac{4}{x-1} + 1$$

$$Y_4 = \frac{4}{x-1} - 1$$

a. In each of the functions, what is the value of a? of b? of c?

b. Which of these values varied?

c. Describe the effect on the graph when this value changed.

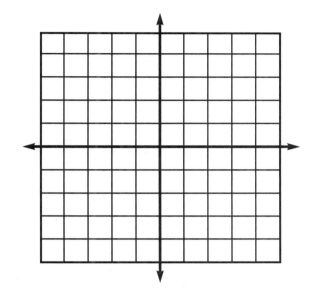

6. Without using the graphing calculator, write your own function of the form $y = \frac{a}{x+b} + c$, where a, b, and c are nonzero values. Sketch the graph, then describe the effect of each constant on your graph. Use the graphing calculator to check your graph.

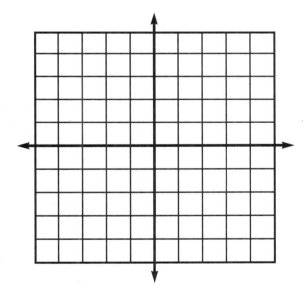

Intersections and Systems of Equations

Level   Algebra I, Algebra II

Using "Intersections" students will

- Enter a system of two equations in y-intercept form
- Use the TABLE menu to generate values
- Use the intersect function to solve the system
- Interpret information from the graph and intersection

Teaching the Lesson

- The TI-82 is required for this lesson.
- Clear all Y= entries.
- Set the TblSet to:
  - TblMin = 0
  - ΔTbl = 1
  - Indpnt: Ask
  - Depend: Ask
- The intersect function is in the CALC menu.

Solutions

1. $G = 30 + 20x$ and $P = 25x$
3. yes
4. At that point, the cost of renting from either company is the same.
5. 6 days
6. Paymore
7. Gotcha
8. Gotcha because their cost is the least after six days of travel.
9. Even though this is the case, if Bob and Dawn rent from another company after 7 days, they will pay the cost of a new rental starting at day 1, not at day eight costs as Bob thinks.
10. $x = -1.6$ $y = 0.2$
11. $x = 0.88$ $y = -2.07$

    the lines are perpendicular
12. $x = 0$ $y = -\frac{2}{3}$

Intersections

Bob Sledd and Dawn Hill are going skiing in Colorado. They plan to rent a car and must choose between Gotcha Rentals and Paymore Rent-a-Car. Gotcha rents cars for $30 down and $20 per day. Paymore rents for just $25 per day.

1. Write an equation to represent the total cost for Gotcha and Paymore in terms of x days.

   $G =$
   $P =$

2. Enter these equations as Y_1 and Y_2. Use the Table function to show the costs of each for the first 10 days.

Paymore		Gotcha	
x	y	x	y
1		1	
2		2	
3		3	
4		4	
5		5	
6		6	
7		7	
8		8	
9		9	
10		10	

Graph these functions using the following WINDOW settings.

$$X_{min} = 0 \qquad Y_{min} = 0$$
$$X_{max} = 10 \qquad Y_{max} = 200$$
$$X_{scl} = 1 \qquad Y_{scl} = 50$$

3. Do these lines intersect?

4. What is true at the intersection of these lines?

Use the intersect function in the $\boxed{\text{CALC}}$ menu to find the intersection of the lines.

5. How many days have passed before they intersect?

6. Which company is least expensive before this time?

7. Which company is most expensive before this time?

8. If Bob and Dawn are going for two weeks, who should they rent from? Why?

9. After Bob saw his graph and chart, he said to Dawn, "Why don't we rent from Paymore the first week and Gotcha the second? That's when they're the cheapest."

Explain the error in his thinking.

10. a. Graph $y = 3x + 5$ and $y = -2x - 3$.

    b. Use the Intersect function to find the solution.

       $x =$ $y =$

11. a. Graph $2x - 3y = 8$ and $-4y = 3 + 6x$.

    b. Use the Intersect function to find the solution.

       $x =$ $y =$

    c. What is special about the two lines? Why?

    Hint : If you're not sure, press ZSquare and look at the graph.

12. a. Graph $5x - 3y = 2$ and $4 + 6y = -10x$.

    b. What is the solution to this system? Why?

Graphing Power © Dale Seymour Publications

Root, Maximum, and Minimum

Level Algebra II

Using "Projectiles and More" students will

- Use the maximum function to find height
- Use the root function to find flight time
- Use the value function to find height at any time
- Use the minimum function

Teaching the Lesson

- The TI-82 is required for this lesson.
- Value, root, maximum, and minimum are found in the $\boxed{\text{CALC}}$ menu.
- The $\boxed{\text{WINDOW}}$ values may need to be adjusted.

Solutions

1a. $s = 100t - (\frac{1}{2})(32)\,t^2$
  b. $s = 156.25$ ft
  c. $t = 6.25$ sec
  d. $s = 136$ ft.
  e. $s = 24$ ft

2a. 601.5 ft
  b. 30.1 sec

3a. $s = 939.8$ ft
  b. $t = 16.4$ sec

4. roots $(-4, 0)$ $(-1, 0)$ minimum $(-2.5, -2.3)$

5. roots $(-0.1, 28.3)$ $(0.4, 0)$ $(0, 2)$
    max $(-0.2, 2.4)$ min $(1.4, -3.8)$

6. roots $(-4, 0)$ $(-2, 0)$ $(2, 0)$ $(3, 0)$
    max $(-0.1, 48.3)$ min $(2.6, -7.4)$ $(-3.2, -31)$

7. the roots match the degree of the equation

8. 5, 6, 12

Projectiles and More

In the 1600s, Galileo accurately described a projectile's motion. He found that the time it takes to go up is the same to come down for a projectile leaving the ground.

This motion can be modeled as a parabola by the following function

$$s = f(t) = vt - \frac{1}{2}gt^2$$

where s is height
            t is time
            g is acceleration due to gravity
            v is the initial velocity

1. Roger Houston is going to test a model rocket. Its initial velocity is 100ft/s

    a. Graph the equation that models this.

    $s =$

    b. Use the maximum function to find the maximum height of the rocket.

    $s =$

    c. Use the root function to find how long it will be airborne.

    $t =$

    d. Use the value function to find how long it takes for the rocket to reach 100 ft.

    $t =$

    e. What is its height after 6 seconds?

Round your answers to questions 2–3 to the nearest tenth.

2. Al Shepard once hit a golf ball on the moon. However, gravity on the moon is $g = 5.32$ ft/s^2. It was quite a shot.

    a. If he threw a moon rock up with an initial velocity of 80 ft/s, how high would it go?

    b. How long would it be before Alan could catch the rock?

Graphing Power © Dale Seymour Publications

3. If Roger Houston took his model rocket to the moon,

   a.   Find the flight time and maximum height.

   b.   What would they be on Mars? ($g = 12.2$ ft/s^2)

   c.   What would they be on Jupiter. ($g = 82.6$ ft/s^2)

Find the roots and max/min values of the following equations.

4. $y = x^2 + 5x + 4$

5. $y = 3x^3 - 5x^2 - 3x + 2$

6. $y = x^4 + x^3 - 16x^2 - 4x + 48$

7. Do you notice a pattern between the roots of these equations and the degree of the equation?

8. How many roots would a fifth degree equation have? Sixth? Twelfth?

Graphing Power © Dale Seymour Publications

TOPIC 6

Geometry and Measurement

Area of a Circle

Level   Pre-Algebra, Informal Geometry, Geometry

Using "Pump Up the Area" the student will

- Input and graph the volume formula for a given radius in a circle
- Observe the effect multiplying the radius by a scale factor has on the area of a circle
- Use TRACE to calculate the area of a circle to complete the table
- Make and test conjectures regarding the area of circles using the graphing capabilities of the graphing calculator

Teaching the Lesson

- When tracing, remind the students that the radius of the circle appears as the x value and the area is the y value.

Solutions

1a. The ratio is the square of the scale factor.
 b. 4

2a. The ratio is the square of the scale factor.
 b. 9

3. 16

4. decreases the area to $\frac{1}{4}$ the original area

Pump Up the Area

Enter the formula for calculating the area of a circle into the $\boxed{Y=}$ menu as $Y_1 = \pi x^2$ and press $\boxed{GRAPH}$.

	TI-81	TI-82
Suggested Range	$X_{min} = -1$	$X_{min} = -1$
	$X_{max} = 94$	$X_{max} = 93$
	$X_{scl} = 2$	$X_{scl} = 2$
	$Y_{min} = -1$	$Y_{min} = -1$
	$Y_{max} = 1259$	$Y_{max} = 1259$
	$Y_{scl} = 10$	$Y_{scl} = 10$
	$X_{res} = 1$	

1. Use $\boxed{TRACE}$ to complete the table.

		Formula	Area	Ratio $\dfrac{A_2}{A_1}$
r	2	$\pi(2)^2$	$A_1 =$	
$2r$	4	$\pi(2 \cdot 2)^2$	$A_2 =$	
r	3		$A_1 =$	
$2r$			$A_2 =$	
r	5		$A_1 =$	
$2r$			$A_2 =$	

a. What do you notice about the ratios?

b. When the radius of a circle is doubled, the area increases by a factor of _____ .

2. Complete the table using the TRACE menu.

		Formula	Area	Ratio $\dfrac{A_2}{A_1}$
r	2	$\pi(2)^2$	$A_1 =$	
$3r$			$A_2 =$	
r	3		$A_1 =$	
$3r$			$A_2 =$	
r	5		$A_1 =$	
$3r$			$A_2 =$	

a. What do you notice about the ratios?

c. When the radius of a circle is tripled, the area increases by a factor of _____.

3. If the radius of a circle is quadrupled, the area increases by a factor of _____. Verify your answer.

4. If the radius of a circle is cut in half, the area of the circle increases/decreases by a factor of _____. Verify your answer.

Graphing Power © Dale Seymour Publications

Volume and Surface Area of Cylinders

Level   Geometry

Using "Cylinders" students will

- Enter a program that calculates a cylinder's volume and its surface area

- Use technology to discover how changing the radius and height of a cylinder affects its volume and surface area

- Use technology to discover the dimensions that give the most or least surface area for a certain volume

Teaching the Lesson

Guide to the keystrokes

Command	Where to Find It
ClrHome	PRGM menu, I/O submenu
Disp	PRGM menu, I/O submenu
to get quote marks	ALPHA , and +
to get a space	ALPHA , and 0
Input	PRGM menu, I/O submenu
Round	MATH menu, NUM submenu
the little arrow	STO menu, EDIT submenu
equal sign	TEST menu

Solutions

5. The volume increases in proportion to height change.

6. The volume changes by the square of the scale factor change in the radius.

7. large radius and short height or very tall with small radius

8. squat cylinders whose radius and height are nearly the same

11. Answers will vary—one solution where $V = 2650$ is $r = 11$, $h = 7$.

Cylinders

The surface area of a cylinder is the measure of the amount of material it takes to wrap the outside of the cylinder.

The volume of a cylinder is the space inside the cylinder that can be filled with something.

1. What units are used to measure the surface area of a cylinder?

2. What units are used to measure the volume of a cylinder?

3. Do you think that the surface area of a cylinder has more, less, or the same number of units as its volume? Why?

The following program will compare the surface area to the volume of a cylinder with any dimensions.

Prgm: CYL

```
:ClrHome
:Disp "COMPARE SURFACE"
:Disp "AREA TO VOLUME"
:Disp "OF A CYLINDER"
:Disp " "
:Disp "RADIUS ="
:Input R
:Disp "HEIGHT ="
:Input H
:Round(2πR(H+R),2)→S
:Round(πR^2H,2)→V
:Disp "SURF. AREA ="
:Disp S
:Disp "VOLUME ="
:Disp V
```

Execute your program using the same radius and height as someone else so that you can compare your solutions. If you both get the same answer, your programs are probably correct.

Graphing Power © Dale Seymour Publications

4. Complete the chart by filling in the missing data. Use the given radii and heights to find surface areas and volumes.

Radius (cm)	Height (cm)	Surface Area (cm^2)	Volume (cm^3)
2	1		
2	2		
2	3		
1	2		
3	2		
3	3		
3	6		
3	9		
3	12		
6	3		
9	3		
12	3		
5.5	3.25		
3.25	5.5		

5. If the radius remains constant but the height changes, how is the volume affected?

6. If the height remains constant but the radius changes, how is the volume affected?

7. In general, which configurations gave the most volume with the least surface area?

8. In general, which configurations gave the most surface area with the least volume?

9. Use the data from the chart to make some generalizations about the relationship of a cylinder's surface area to its volume.

10. A cylinder has a radius of 3 cm and a height of 15 cm.

    a. Estimate its surface area.

    b. Estimate its volume.

    c. Calculate its actual surface area and volume.

11. You are working for a company that makes cylindrical containers. The government has hired your company to manufacture a cylinder that they require to hold approximately 2650 cubic cm of nuclear waste (the container can hold up to 10 cm^3 more than 2650 cm^3, or it can hold as little as 2640 cm^3). The material that you must use to make the containers is very expensive, so your employer wants you to design a container that meets the government's requirement for volume while using as little material (surface area) as possible to construct each container.

    What general cylindrical shape do you predict will be the best for this problem? (Examples: tall and thin like a potato chip can; about as tall as it is wide, like the top of a spray paint can; or short and wide, like a tuna can.)

    Try cylinders of several dimensions; keep track of the numbers you use. When you feel that you have found the best possible container, draw a diagram of your container showing all the important information.

Graphing Power © Dale Seymour Publications

Volume of a Sphere

Level   Informal Geometry, Geometry

Using "Pump Up the Volume" the student will

- Use the graphing capabilities to graph the volume of a sphere
- Make and test conjectures to discover how the radius affects the volume of a sphere
- Calculate the volume of a sphere given the radius

Teaching the Lesson

π is a character on the calculator, if you prefer, the students may use the approximation of 3.14. Using the MODE menu, you may set the number of digits the calculator will display.

Solutions

  3a.   ratio is 8
    b. 2^3
    c.   8

  4a.   ratio is 27
    b. 3^3
    c.   27

  5a.   64
    b.   decreases to $\frac{1}{8}$ the original volume

Pump Up the Volume

1. In the $\boxed{Y=}$ menu, enter $Y_1 = (\frac{4}{3})\pi r^3$.

2. Suggested Range

	TI-81	TI-82
	$X_{min} = -2.7$	$X_{min} = -2.6$
	$X_{max} = 6.8$	$X_{max} = 6.8$
	$X_{scl} = 1$	$X_{scl} = 1$
	$Y_{min} = -1000$	$Y_{min} = -1000$
	$Y_{max} = 5300$	$Y_{max} = 5300$
	$Y_{scl} = 100$	$Y_{scl} = 100$
	$X_{res} = 1$	

3. Complete the following using the $\boxed{\text{TRACE}}$ function.

		Formula	Area	Ratio $\dfrac{V_2}{V_1}$
r	2	$\frac{4}{3}\pi(2)^3$	$V_1 =$	
$2r$			$V_2 =$	
r	3		$V_1 =$	
$2r$			$V_2 =$	
r	5		$V_1 =$	
$2r$			$V_2 =$	

    a. What do you notice about the ratios?

    b. How is this value related to 2?

    c. When the radius of a sphere is doubled, the volume increases by a factor of
    _____.

Graphing Power © Dale Seymour Publications

4. Complete the following using the $\boxed{\text{TRACE}}$ function.

		Formula	Area	Ratio $\dfrac{V_2}{V_1}$
r	2	$\dfrac{4}{3}\pi(2)^3$	$V_1 =$	
$3r$			$V_2 =$	
r	3		$V_1 =$	
$3r$			$V_2 =$	
r	5		$V_1 =$	
$3r$			$V_2 =$	

    a.   What do you notice about the ratios?

    b.   How is this value related to 3?

    c.   When the radius of a sphere is tripled, the volume increases by a factor of

            _____.

5. If the radius of a sphere is quadrupled, the volume increases by a factor of _____. Verify your answer.

6. If the radius of a sphere is halved, the volume increases/decreases by a factor of _____. Verify your answer.

Distances, Slopes, Midpoints, and Geometric Figures

Level   Geometry, Algebra II, Analytic Geometry

Using "Points, Segments, and Figures" the student will

- Use the programming capabilities to calculate distance, midpoint, and slope, given two points

- Use the program to determine if a set of points are the vertices of a particular geometric figure

Teaching the Lesson

- This lesson involves a substantial amount of programming. Use only if your class has had practice inputting and debugging programs.

- Mastery in calculating and using distance, midpoint, slope, intercept, and slope-intercept formulas is assumed in this lesson.

- The programs in this lesson deal with points plotted on a plane, the distance formula, the midpoint between two points, and the slope formula, as well as the slope-intercept form of a line. The programming feature of the graphing calculator will be used extensively.

Points, Segments, and Figures

1. For any two points (x_1, y_1) and (x_2, y_2)

   a. How do you find the distance between them?

   b. How do you find their midpoint?

   c. How do you find the slope of the line that contains them?

   d. How do you find the y-intercept of the line that contains them?

2. Using the points $(-3, 4)$ and $(3, -2)$

   a. Find the distance between them.

   b. Find the midpoint.

   c. Find the slope of the line that contains them.

   d. Find the y-intercept of the line that contains them.

   e. Write the equation of the line in the form $y = ax + b$.

3. Enter the following program into your calculator. Title the program DISTFORM. An explanation of each step is provided. If you are using the TI-82, replace :End with :Stop.

Program Steps	Explanation
:Disp "X1="	display what is in quotes
:Input Q	stores your value of x_1 in Q
:Disp "Y1="	displays what is in quotes
:Input R	stores your value of y_1 in R
:Disp "X2="	displays what is in quotes
:Input S	stores your value of x_2 in S
:Disp "Y2="	displays what is in quotes
:Input T	stores your value of y_2 in T
:$\sqrt{((Q-S)^2+(R-T)^2)} \rightarrow D$	computes the distance formula and stores in D
:Pause	pauses; pressing $\boxed{\text{ENTER}}$ continues
:Disp "DIST ="	displays what is in quotes
:Disp D	displays the value in D

Run the program using the points $(-3, 4)$ and $(3, -2)$ to be sure the distance matches problem 2.

4. Use DISTFORM to find the distance between each pair of points

    a. $(4, 2), (6, 6)$ c. $(7, -9), (7, -1)$

    b. $(\frac{1}{2}, \frac{9}{2}), (-2, -\frac{3}{2})$ d. $(4.8, 2.2), (4.8, -2.8)$

5. Edit DISTFORM by adding these lines to the end of the program.

Program Steps	Explanation
`:(Q+S)/2→M`	computes the x coordinate of the midpoint and stores in M
`:Disp "MIDPT(X)="`	displays what is in quotes
`:Disp M`	displays the value in M
`:(R+T)/2→N`	computes the Y coordinate of the midpoint and stores in N
`:Disp "MIDPT(Y)="`	displays what is in quotes
`:Disp N`	displays the value in N
`:If (Q-S)=0`	Checks for a vertical line
`:Goto 2`	jumps to the '2' label
`:(R-T)/(Q-S)→A`	calculates the slope and stores in A
`:Disp "SLOPE="`	displays what is in quotes
`:Disp A`	displays the value in A
`:End`	ends program
`:Lbl 2`	labels this line as '2'
`:Disp "SLOPE UNDEF"`	displays what is in quotes

6. Which lines of the program account for an undefined slope? How does it do so?

7. Run the revised DISTFORM program to find the coordinates of the midpoint and the slope of the line going through the points $(-3, 4)$ and $(3, -2)$. Compare it to the answer you got in problem 2, then run it to find the midpoint and the slope for each ordered pairs.

    a. $(4, 2), (6, 6)$ c. $(7, -9), (7, -1)$

    b. $(\frac{1}{2}, \frac{9}{2}), (-2, -\frac{3}{2})$ d. $(4.8, 2.2), (4.8, -2.8)$

8. Now edit the DISTFORM program so it looks like this

Graphing Power © Dale Seymour Publications

Program Steps	Explanation
:Lbl 1	labels this line as '1'
:Disp "X1="	displays what is in quotes
:Input Q	stores your value of x_1 in Q
:Disp "Y1="	displays what is in quotes
:Input R	stores your value of y_1 in R
:Disp "X2="	displays what is in quotes
:Input S	stores your value of x_2 in S
:Disp "Y2="	displays what is in quotes
:Input T	stores your value of y_2 in T
:√((Q-S)2+(R-T)2)→D	computes the distance formula and stores in D
:Pause	pauses; pressing ENTER continues
:Disp "DIST ="	displays what is in quotes
:Disp D	displays the value in D
:Pause	pauses; pressing ENTER continues
:(Q+S)/2→M	computes the x coordinate of the midpoint and stores in M
:Disp "MIDPT(X)="	displays what is in quotes
:Disp M	displays the value in M
:(R+T)/2→N	computes the y coordinate of the midpoint and stores in N
:Disp "MIDPT(Y)="	displays what is in quotes
:Disp N	displays the value in N
:If (Q-S)=0	Checks for a vertical line
:Goto 2	jumps to the '2' label
:(R-T)/(Q-S)→A	calculates the slope and stores in A
:Disp "SLOPE="	displays what is in quotes
:Disp A	displays the value in A
:R-AQ→B	computes the y-intercept and stores in B
:Disp "Y INT="	displays what is in quotes
:Disp B	displays the value in B
:End	ends program
:Lbl 2	labels this line as '2'
:Disp "SLOPE UNDEF"	displays what is in quotes
:Disp "Y INT-NONE"	displays what is in quotes

To repeat the program over without having to QUIT and start again,

- change the END to a Goto 1.

- add Goto 1 to the end of the program.

9. Use your revised DISTFORM program to find the equation of the line going through each pair of points. Write your equations in the form of $y = ax + b$.

   a. $(4, 2), (6, 6)$ c. $(7, -9), (7, -1)$

   b. $(\frac{1}{2}, \frac{9}{2}), (-2, -\frac{3}{2})$ d. $(4.8, 2.2), (4.8, -2.8)$

   Use DISTFORM to answer the following questions. Justify each answer in a clear and informative manner.

10. What type of quadrilateral has its vertices at $(1, 7), (3, 5), (4, -1), (2, 1)$?

11. What type of quadrilateral has its vertices at $(1, 1), (4, 4), (2, 6), (-1, 3)$?

12. What type of quadrilateral has its vertices at $(-6, 3), (-1, 6), (2, 1), (-3, -2)$?

13. What type of quadrilateral has its vertices at $(0, 1), (2, -3), (-2, -1), (-4, 3)$?

14. What type of quadrilateral has its vertices at $(2, 0), (4, -6), (9, 1), (7, 7)$?

Graphing Power © Dale Seymour Publications

15. What type of quadrilateral has its vertices at $(7, 4)$, $(10, 1)$, $(6, -2)$, $(-1, -2)$?

16. Show that $(-3, 3)$, $(1, 11)$, and $(3, 15)$ are collinear.

17. Show that $(-1, 2)$, $(-6, -2)$, $(2, -12)$ are the vertices of a right triangle.

18. Three vertices of a parallelogram have coordinates $(-3, 1)$, $(1, 4)$, $(4, 3)$. Find the coordinates of the fourth vertex.

19. Modify DISTFORM to draw the geometric figures formed by your points. Insert the following line into your program after you enter the values for x_1, y_1, x_2, y_2, the Line command is in the ⟦DRAW⟧ menu. It draws the line between (Q, R) and (S, T). You need to do this for each correct pair of points.

:Line (Q,R,S,T)

Area Under a Curve

Level   Geometry, Algebra II, Precalculus, Calculus

Using "Fiesta City Roller Coaster" students will

- Calculate the area under a curve

- Use the graphing capabilities to compare curves

Teaching the Lesson

- Depending on the experience and ability of the students, you may need to create several problems in which they find the area bounded by the axes and horizontal and vertical lines.

- Have students find and record the values of the function $y = x^2$ at the points $x = \{0, 3, 6, 9, 12, 15, 18\}$ and discuss how rectangles or trapezoids can be used to find the area under the curve.

- Use the calculator to find the area of the trapezoid between $x = 6$, $x = 9$, the x-axis, and an approximation of $y = x^2$.

Solutions

1. 1944 ft^2

2. $\$21,600$

3. $26,244 \text{ ft}^2$; $\$180,500$

4. 3642 ft^2

5. 4436 ft^2

Fiesta City Roller Coaster

The manager of the Fiesta City Amusement Park in San Antonio would like to cover the north slope of the roller coaster with a large banner announcing the name of the roller coaster for opening day.

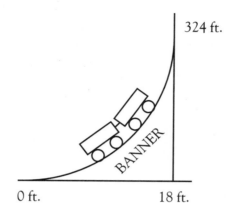

324 ft.

0 ft.   18 ft.

Answer the following questions using the graphing calculator.

1. If the shape of the roller coaster closely resembles a parabola $y = x^2$, what is the area that will be covered by the banner? (Show work and explain your reasoning.)

2. If the San Antonio Flag Company were commissioned to make the banner and they charged $100.00 per square yard, how much will the banner cost?

3. If the shape of the roller coaster was more like the function $y = x^3$, what would the area and the cost be?

Graphing Power © Dale Seymour Publications

4. If the roller coaster was shaped more like $y = 1.5x$, what would the area be?

5. If the roller coaster was shaped more like $y = 400 \sin (0.0826x)$, what would the area be?

6. If you love the fast speed at which a roller caster goes down a hill, which of the following roller coasters would you like to ride and why?

   A Fiesta City Roller Coaster with an equation of

   a. $y = x^2$ b. $y = 1.5x$ c. $y = 400 \sin (0.0826x)$

Graphing Power © Dale Seymour Publications

TOPIC 7

Conics

Conic Sections

Level   Algebra II, Precalculus

Using "Graphs of Conics" the students will

- Graph various conic equations on the graphing calculator

- Make and test conjectures regarding the effect changing a constant has on the conic section and its graph

Teaching the Lesson

- Prior to this activity, students should be familiar with conic sections and their equations.

- Programs (from *TI-81 Newsletter*, Volume 2, February, 1992) are included for convenience. Your program numbers may be different from the numbers provided. Remember, if program 1 is, for you, program 13, you must change *every* occurrence of Prgm1 to Prgm13.

- This lesson is designed for students who have experience in programming. It will take them more than an hour to enter these six programs. Students working in groups can share the task of entering the programs into one calculator.

- These programs are designed to be run by the control program (Prgm 6:CONICSEC). If you want to run them individually, they will need to be modified slightly.

Graphs of Conics

Circle: $(x - h)^2 + (y - k)^2 = r^2$

Run the main program and select the circle option. Use the following as values requested by the calculator.

1. Let $h = 0$, $k = 0$, and $r = 3$. Let $h = 0$, $k = 0$, and $r = 4$.

   What effect does a change in r have on the graph?

2. Let $h = 1$, $k = 0$, and $r = 3$. Let $h = -1$, $k = 0$, and $r = 3$.

   What effect does a change in h have on the graph?

3. Let $h = 0$, $k = 1$, and $r = 3$. Let $h = 0$, $k = -1$, and $r = 4$.

   What effect does a change in k have on the graph?

4. Let $h = 0$, $k = 0$, and $r = 9$. Let $h = 8$, $k = 4$, and $r = 3$.

   What effect did changing the values of h, k, and r have? What limitations does the program have? How can you adjust for these limitations?

Parabola: $y = a(x - h)^2 + k$ and $x = a(y - k)^2 + h$

Run the main program and select the parabola option. Use the following as values requested by the calculator.

Run problem 5 twice using both the $x =$ and the $y =$ form of the equation.

5. Let $h = 0$, $k = 0$, and $a = 1$.

   How did each differ and why?

Run 6–8 using the y = form of the equation

6. Let $h = 0$, $k = 0$, and $a = 1$.
   Let $h = 0$, $k = 0$, and $a = -1$.

   What effect does changing the sign of a have on the graph?

7. Let $h = 0$, $k = 1$, and $a = 1$. Let $h = 0$, $k = 1$, and $a = 2$.

   What effect does changing a have on the graph?

8. Let $h = 1$, $k = 0$, and $a = 1$. Let $h = -1$, $k = 0$, and $a = 1$. Let $h = 0$, $k = 1$, and $a = 1$. Let $h = 0$, $k = -1$, and $a = 1$.

   What effect does changing h and/or k have on the graph?

Repeat 6–8 using the x = form of the equation.

$$\text{Ellipse: } 1 = \frac{(x - h)^2}{a^2} + \frac{(y - k)^2}{b^2}$$

Run the main program and select the ellipse option. Use the following as values requested by the calculator.

9. Let $h = 0$, $k = 0$, $a = 3$, and $b = 2$.
   Let $h = 0$, $k = 0$, $a = 2$, and $b = 3$.

   What effect do a and b have on the graph?

10. Let $h = 1$, $k = 0$, $a = 3$, and $b = 2$.
    Let $h = -1$, $k = 0$, $a = 3$, and $b = 2$.

    What effect does a change in h have on the graph?

11. Let $h = 0$, $k = 1$, $a = 3$, and $b = 2$.
    Let $h = 0$, $k = -1$, $a = 3$, and $b = 2$.

Graphing Power © Dale Seymour Publications

What effect does a change in k have on the graph?

12. Let $h = 0$, $k = 0$, $a = 3$, and $b = 3$.
What is familiar about this graph? Why is this?

$$\text{Hyperbola: } 1 = \frac{(x - h)^2}{a^2} - \frac{(y - k)^2}{b^2} \text{ or } 1 = \frac{(y - k)^2}{b^2} - \frac{(x - h)^2}{a^2}$$

Run the main program and select the hyperbola option. Use the following as values requested by the calculator.

Run problem 13 using both forms of the equation.

13. Let $h = 0$, $k = 0$, $a = 1$, and $b = 1$.

How did each differ and why?

Run 14–16 using the first form of the equation

14. Let $h = 0$, $k = 0$, $a = 3$, and $b = 2$.
Let $h = 0$, $k = 0$, $a = 2$, and $b = 3$.

What effect do a and b have on the graph?

15. Let $h = 1$, $k = 0$, $a = 3$, and $b = 2$.
Let $h = -1$, $k = 0$, $a = 3$, and $b = 2$.

What effect does a change in h have on the graph?

16. Let $h = 0$, $k = 1$, $a = 3$, and $b = 2$.
Let $h = 0$, $k = -1$, $a = 3$, and $b = 2$.

What effect does a change in k have on the graph?

Repeat 14–16 using the second form of the equation.

Programs for Use with Conic Sections

TI-81

```
:Prgm1: RNGSET
:-9.6→Xmin
:9.4→Xmax
:1→Xscl
:-6.4→Ymin
:6.2→Ymax
:1→Yscl


:Prgm2: CIRCLE
:Prgm1
:ClrHome
:Disp "(X-H)²+(Y-K)²=R²"
:Disp "H"
:Input H
:Disp "K"
:Input K
:Disp "R"
:Input R
:"√(R²-(X-H)²)+K"→Y1
:"-(Y1-2K)"→Y2
:PT-On(H,K)


Prgm3: PARABOLA
:Prgm1
:ClrHome
:Disp "1) Y=A(X-H)²+K"
:Disp "2) X=A(Y-K)²+H"
:Disp "CHOOSE 1 OR 2"
:Input J
:Disp "H"
:Input H
:Disp "K"
:Input K
:Disp "A"
:Input A
:If J=2
:Goto X
:"A(X-H)²+K"→Y1
:H→M
:K+(4A)⁻¹→N
:Goto G
:Lbl X
:"√((X-H)/A)+K"->Y1
:H+(4A)⁻¹→M
```

TI-82

```
:ClrDraw:AxesOn
:-9.6→Xmin
:9.4→Xmax
:1→Xscl
:-6.4→Ymin
:6.2→Ymax
:1→Yscl
:Lbl B
:Menu("CONICS
    GRAPHER","CIRCLE",M,"PARABOLA",N,
    "ELLIPSE",P,"HYPERBOLA,"Q,"QUIT?",
    V)
:Lbl M
:ClrHome
:Output(2,1,"(X-H)²+(Y-K)²=R²")
:Input"H",H
:Input"K",K
:Input"R",R
:"√(R²-(X-H)²)+K"→Y1
:"-(Y1-2K)"→Y2
:Pt-On(H,K)
:DispGraph:Pause
:Goto B
:Lbl N
:Menu("CHOOSE
    PARABOLA","Y=A(X-H)²+K",R,"X=A(Y-K
    )²+H",S)
:Lbl R
:Input"H",H
:Input"K",K
:Input"A",A
:A(X-H)²+K→Y1
:H→M
:K+(4A)⁻¹"→N
:DispGraph:Pause
:Pt-On(M,N)
:Text(1,1,"DIRECTRIX",M,",",N)
:Pause
:"K-(4A)⁻¹"→Y3
:DispGraph
:Pause
:Goto B
:Lbl S
:Input"H",H
```

Graphing Power © Dale Seymour Publications

TI-81 (cont.)

```
:K->N
:"-(Y₁-2K)"→Y₂
:Lbl G
:DispGraph
:Pause
:Disp "SHOW DIRECTRIX, PRESS ENTER"
:Pause
:"K-(4A)⁻¹"→Y₃
:IF J=2
:"10000(X-(H-(4A)⁻¹))"→Y₃
:PT-On(M,N)

:Prgm4: ELLIPSE
:Prgm1
:ClrHome
:Disp  "(X-H)²        (Y-K)²"
:Disp "—-  +   —- = 1"
:Disp " A²              B²"
:Disp "H"
:Input H
:Disp "K"
:Input K
:Disp "A"
:Input A
:Disp "B"
:Input B
:"(√(B²(1-(X-H)²/A²)))+K"→Y₁
:"-(Y₁-2K)"→Y₂
:PT-On (H,K)

:Prgm5: HYPERBOLA
:Prgm1
:ClrHome
:Disp "(X-H)²        (Y-K)²"
:Disp "—-   -   —-=1"
:Disp "  A²              B²"
:Disp "(Y-K)²        (X-H)²"
:Disp "—-   -   —-=1"
:Disp " B²              A²"
:Disp "1=TOP 2=BOTTOM"
:Input J
:Disp "H"
:Input H
:Disp "K"
:Input K
:Disp "A"
```
```

## TI-82 (cont.)

```
:Input"K",K
:Input"A",A
"√((X-H)/A)+K"→Y₁
:H+(4A)⁻¹→M
:K→N
:"-(Y₁-2K)"→Y₂
:"10000(X-(H-(4A)⁻¹))→Y₃
:DispGraph
:Text(1,1,"DIRECTRIX",M,",",N)
"Pt-On(M,N)
:Pause
:Goto B
:Lbl P
:ClrHome
:Disp"(X-H)²  (Y-K)²"
:Disp "—-  -  —-=1"
:Disp "  A²        B²"
:Input"H",H
:Input"K",K
:Input"A",A
:Input"B",B
:"(√(B²(1-(X-H)²/A²)))+K→Y₁
:"-(Y₁-2K)"→Y₂
:Pt-On(H,K)
:DispGraph
:Goto B
:Lbl Q
:ClrHome
:Disp "(X-H)²        (Y-K)²"
:Disp "—-   -   —-=1"
:Disp "  A²            B²"
:Disp "(Y-K)²        (X-H)²"
:Disp "—-   -   —-=1"
:Disp "  B²            A²"
:Disp "1=TOP 2=BOTTOM"
:Input J
:Disp "H"
:Input H
:Disp "K"
:Input K
:Disp "A"
:Input A
:Disp "B"
:Input B
:If J=2
```

Graphing Power © Dale Seymour Publications
```

TI-81 (cont.)

```
:Input A
:Disp "B"
:Input B
:If J=2
:Goto X
:"(√(B²((X-H)²/A²-1)))+K"→Y₁
:"-(Y₁-2K)"→Y₂
:Goto G
:Lbl X
:"(√(B²((X-H)²/A²+1)))+K"→Y₁
:"-(Y₁-2K)"→Y₂
:Lbl G
:Input
:Disp "SHOW ASYMPTOTES, PRESS ENTER."
:Pause
:"(B/A)(X-H)+K"→Y₃
:"-(Y₃-2K)"→Y₄

:Prgm6: CONICSEC
:Lbl B
:Disp "CONIC SECTION"
:Disp "GRAPHER. CHOOSE"
:Disp "CIRCLE   =1"
:Disp "PARABOLA =2"
:Disp "ELLIPSE  =3"
:Disp "HYPEPBOLA=4"
:Disp "FINISHED =5"
:Input I
:If I=5
:Goto Z
:""->Y₂
:""->Y₃
:""->Y₄
:If I=1
:Prgm2
:If I=2
:Prgm3
:If I=3
:Prgm4
:If I=4
:Prgm5
:DispGraph
:Input
:Goto B
:Lbl Z
:DispGraph
```

TI-82 (cont.)

```
:Goto X
:"(√(B²((X-H)²/A²-1)))+K"→Y₁
:"-(Y₁-2K)"→Y₂
:Goto G
:Lbl X
:"(√(B²((X-H)²/A²+1)))+K"→Y₁
:"-(Y₁-2K)"→Y₂
:Lbl G
:Input
:Disp "SHOW ASYMPTOTES, PRESS ENTER."
:Pause
:"(B/A)(X-H)+K"→Y₃
:"-(Y₃-2K)"→Y₄
:DispGraph
:Goto B
:Lbl V
:Stop
```

Graphing Power © Dale Seymour Publications

Graphs of Conic Sections

Level   Algebra II, Precalculus

Using "Conics" the student will

- Use the graphing capabilities to make and test conjectures about conic sections and their graphs

- Enter functions and observe the differences and similarities among the graphs

- Describe in writing how the constants in each function change the graph

Teaching the Lesson

- All equations must be solved for y before entering them into the $\boxed{\text{Y=}}$ menu. The calculator will graph functions only, so both halves of the equation must be entered into the $\boxed{\text{Y=}}$ menu separately. For ease and speed in entering the two halves of the equation, use the $\boxed{\text{Y-VARS}}$ menu to enter $\boxed{(-)}$ Y_1 for Y_2.

- The $\boxed{\text{ZOOM}}$ menu allows you to reset the range to Standard or ZSquare. The square window is important when graphing circles.

- To graph only one equation at a time, deselect the other equations by moving the cursor over the = sign and pressing $\boxed{\text{ENTER}}$. The = sign will no longer be highlighted and will not be graphed. To reselect the equation move the cursor back over the = sign and press $\boxed{\text{ENTER}}$. The equation will again be graphed.

Conics

Enter each set of equations in the $\boxed{\text{Y=}}$ menu of the calculator and graph them one at a time on the same graph grid. Draw a sketch of the conic on the graph grid and answer the questions.

1. Graph.
$$9x^2 + 4y^2 = 36$$
$$4x^2 + 4y^2 = 36$$
$$9x^2 - 4y^2 = 36$$
$$9x^2 + 4y = 36$$

   a. Compare the equation with the graph. What effect do signs and coefficients have on the graphs?

   b. How do they differ?

   c. What do they have in common?

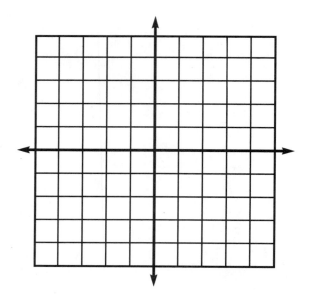

2. Graph each circle.
$$x^2 + y^2 = 16$$
$$x^2 + y^2 = 25$$
$$x^2 + y^2 = 20$$
$$x^2 + y^2 = 0$$

   a. What is the center and radius of each circle?

   b. Why is $x^2 + y^2 = 0$ degenerate?

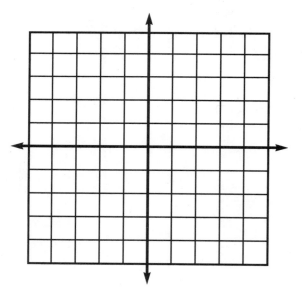

Graphing Power © Dale Seymour Publications

3. Graph each circle.

$(x - 3)^2 + (y + 4)^2 = 16$

$(x + 2)^2 + (y - 1)^2 = 25$

$(x - 1)^2 + y^2 = 20$

$x^2 + y^2 - 2x + 6y = -1$

(Change $x^2 + y^2 - 2x + 6y = -1$ to standard form by completing the squares.

$(x^2 - 2x + 1) + (y^2 + 6y + 9) = -1 + 10)$

a. What is the center and radius of each circle?

b. How do $x^2 + y^2 = 16$ and $(x - 3)^2 + (y + 4)^2 = 16$ compare?

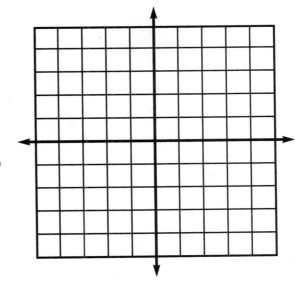

4. Graph each ellipse.

$\dfrac{x^2}{25} + \dfrac{y^2}{4} = 1$

$\dfrac{x^2}{4} + \dfrac{y^2}{25} = 1$

$x^2 + 4y^2 = 16$

$x^2 + 4y^2 = 0$

a. List the major axis.

b. Why is $x^2 + 4y^2 = 0$ degenerate?

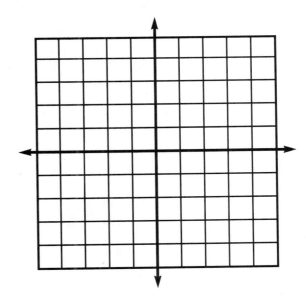

5. Graph each ellipse.

$$\frac{(x + 3)^2}{25} + \frac{(y - 2)^2}{4} = 1$$

$$\frac{(x - 3)^2}{4} + \frac{(y + 2)^2}{25} = 1$$

$$x^2 + 4y^2 - 2x + 16y = -1$$

a. List the center of each.

b. List the major axis.

c. How do $\dfrac{x^2}{25} + \dfrac{y^2}{4} = 1$ and

$\dfrac{(x + 3)^2}{25} + \dfrac{(y - 2)^2}{4} = 1$

compare?

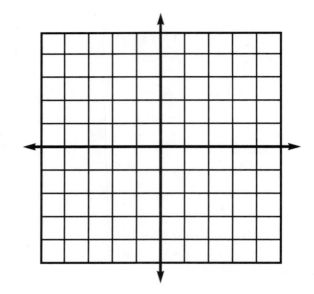

6. Graph each hyperbola.

$$\frac{x^2}{25} - \frac{y^2}{4} = 1$$

$$\frac{y^2}{4} - \frac{x^2}{25} = 1$$

$$x^2 - 4y^2 = 16$$

$$y^2 - 4x^2 = 0$$

a. List the major axis.

b. Why is $y^2 - 4x^2 = 0$ degenerate?

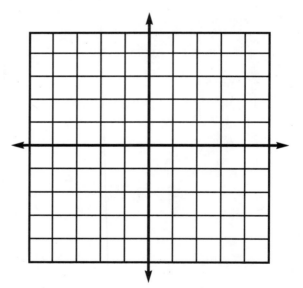

Graphing Power © Dale Seymour Publications

7. Graph.

$$\frac{(x+3)^2}{25} - \frac{(y-2)^2}{4} = 1$$

$$\frac{(y-3)^2}{4} - \frac{(y+2)^2}{25} = 1$$

$$x^2 - 4y^2 - 2x + 16y = 19$$

a. List the center of each.

b. List the x and y intercepts.

c. How do $\dfrac{x^2}{25} + \dfrac{y^2}{4} = 1$ and

   $$\frac{(x+3)^2}{25} + \frac{(y-2)^2}{4} = 1$$
   compare?

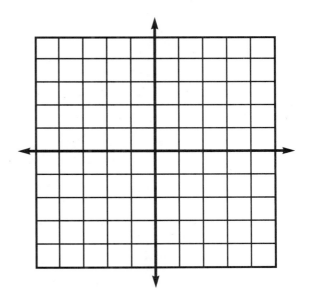

8. Graph each parabola.

$$y - x^2 = 2x - 3$$
$$x^2 + y = 4x - 3$$
$$x - y^2 = 2y - 3$$
$$x + y^2 = 4y - 3$$

a. Explain the affects of the squared variable.

b. Using the general form, $a(x-k)^2 + h = y$, explain the effect the sign of a has on the graph.

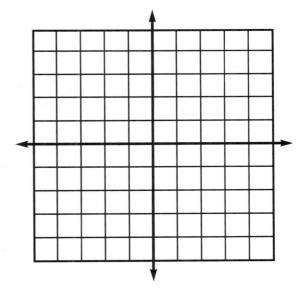

TOPIC 8

Precalculus

Logarithms

Level   Algebra II, Precalculus

Using "Properties of Logarithms" students will

- Evaluate logarithms using the graphing calculator
- Identify relationships of logarithms within each problem set
- Use results to generalize the properties of logarithms

Teaching the Lesson

- Set the ⎡MODE⎤ menu to 1 decimal place before beginning the lesson.

Solutions

1. To find the log of a product, add the log values.

2. To find the log of a quotient, subtract the log values.

3. To find the log of a power, multiply the log value by the exponent.

Properties of Logarithms

1. Log of a Product

   a. Find

   log 100 = _____
   log 1000 = _____
   log (100 · 1000) = _____

   b. Find

   log 5 = _____
   log 7 = _____
   log (5 · 7) = _____

   c. Find

   log 20 = _____
   log 8 = _____
   log (20 · 8) = _____

   d. Use the solutions to 1a, 1b, and 1c to state the relationship of the third log expression to the first and second log expressions.

2. Log of a Quotient

   a. Find

   log 1000 = _____
   log 10 = _____
   log (1000 ÷ 10) = _____

   b. Find

   log 18 = _____
   log 9 = _____
   log (18 ÷ 9) = _____

   c. Find

   log 55 = _____
   log 15 = _____
   log (55 ÷ 15) = _____

   d. Use the solutions to 2a, 2b, and 2c to state the relationship of the third log expression to the first and second log expressions.

3. Log of a Power

    a. Find

$$\log 1000 = \text{_____}$$
$$\log (1000)^2 = \text{_____}$$

    b. Find

$$\log 33 = \text{_____}$$
$$\log (33)^3 = \text{_____}$$

    c. Find

$$\log 25 = \text{_____}$$
$$\log (25)^4 = \text{_____}$$

    d. Use the solutions to 3a, 3b, and 3c to state the relationship of the second log expression to the first log expression.

4. Write a general rule (property) for each relationship.

Question 1 $\log (x \cdot y) \quad =$

Question 2 $\log (x \div y) \quad =$

Question 3 $\log (x)y \quad =$

Graphing Power © Dale Seymour Publications

3. $y = \tan x$

   a. Where does the curve cross the
   x-axis?

   b. At what values of x is the
   function a maximum or a
   minimum?

   c. What is the maximum value of
   the function?

   d. What is the minimum value of
   the function?

   e. Is there a problem with
   answering questions 3b–3d?
   Why or why not?

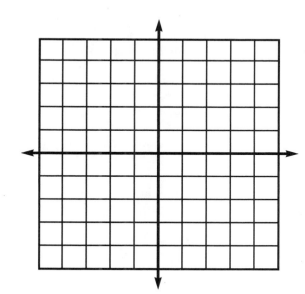

4. $y = \cot x$
[remember $\cot x = (\tan x)^{-1}$]

   a. Where does the curve cross the
   x-axis?

   b. At what values of x is the
   function a maximum or a
   minimum?

   c. What is the maximum value of
   the function?

   d. What is the minimum value of
   the function?

   e. Do you have the same problem
   with answering questions
   4b–4d as you did with the
   graph of $y = \tan x$? At the same
   place?

   f. Graph $y = \tan x$ without clear-
   ing $y = \cot x$. What generalizations can you make about the
   differences between the
   graphs? (Look for patterns.)

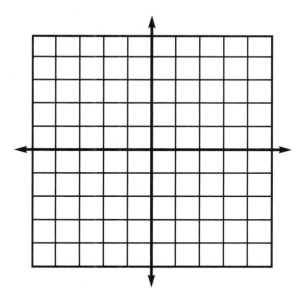

5. $y = \sec x$

  a. Where does the curve cross the
     x-axis?

  b. At what values of x is the
     function a maximum or a
     minimum?

  c. What is the maximum value of
     the function?

  d. What is the minimum value of
     the function?

  e. For which questions do there
     seem to be no answer? Why?

  f. Graph $y = \cos x$ without clear-
     ing $y = \sec x$. What generaliza-
     tions can you make about the
     differences between the
     graphs? (Look for patterns.)

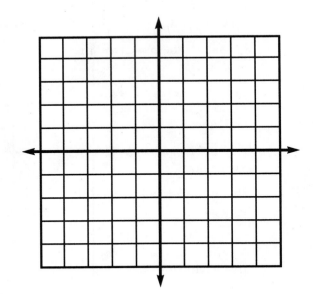

6. $y = \csc x$

  a. Where does the curve cross the
     x-axis?

  b. At what values of x is the
     function a maximum or a
     minimum?

  c. What is the maximum value of
     the function?

  d. What is the minimum value of
     the function?

  e. Do you have the same problem
     with answering questions
     6a–6d as you did with the
     graph of $y = \sec x$? Do the
     problems come at the same
     place?

  f. Graph $y = \sin x$ without clear-
     ing $y = \csc x$. What generaliza-
     tions can you make about the
     differences between the
     graphs? (Look for patterns.)

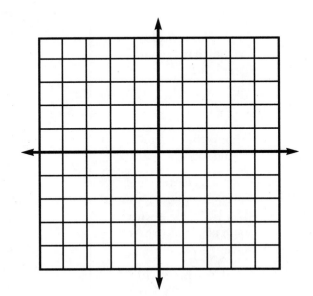

Graphing Power © Dale Seymour Publications

7. Graph $y = \sin x$ and $y = \cos x$ on the same screen. What generalizations can you make about the differences between the graphs?

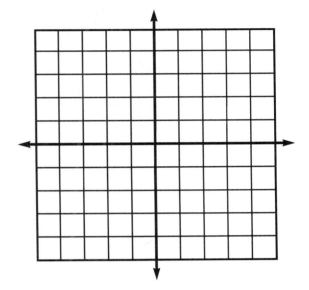

8. If you were to graph $y = \sec x$ and $y = \csc x$ on the same screen, what pattern would you see?

Graphing Power © Dale Seymour Publications

Sinusoidal Functions

Level   Analytical Geometry, Precalculus

Using "New Wave" the student will

- Use the graphing capabilities of the calculator to graph sine functions

- Observe the differences and similarities among the graphs of sine functions

- Predict and test the effects of changing A, B, C, and D on the graph of $y = A \sin B(x - C) + D$

- Describe in writing, how the constants in the sinusoidal equation of $y = A \sin B(x - C) + D$ affect the graph of $y = \sin x$.

Teaching the Lesson

- To use the trig screen, go to the $\boxed{\text{MODE}}$ menu and select Degree. Use the $\boxed{\text{ZOOM}}$ menu to select ZTrig. Set your range.
  $X_{min} = -360$, $X_{max} = 360$, $X_{scl} = 90$, $Y_{min} = -5$, $Y_{max} = 5$, $Y_{scl} = 1$.

- Students need to clear the $\boxed{\text{Y=}}$ menu before starting each new set of trigonometric functions.

- Warning:
  sin 4(A + B) evaluates as (sin 4)(A + B)
  sin (4(A + B)) or sin (A + B)4 evaluates as sin 4(A + B)

New Wave

Enter each set of equations into the $\boxed{Y=}$ menu of the calculator and $\boxed{GRAPH}$ them one at a time on the same graph grid. Draw a sketch of the calculator display on the graph grid and answer the questions.

1.

$Y_1 = \sin x$
$Y_2 = 2 \sin x$
$Y_3 = 3 \sin x$
$Y_4 = 4 \sin x$

a. How are these curves alike? Different?

b. What happens to the graph of $y = \sin x$ when $y = A \sin x$ where $A > 1$?

c. Use the calculator to explore $y = A \sin x$ where $0 < A < 1$. What happens?

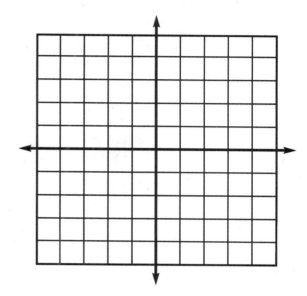

2.

$Y_1 = \sin x$
$Y_2 = -\sin x$

a. How are these curves alike? Different?

b. What happens to the graph of $y = \sin x$ when $y = A \sin x$ where $A < 0$?

c. What affect does a negative coefficient have on the graph of $y = \sin x$?

d. Describe how a coefficient of -2 will affect the graph of $y = \sin x$.

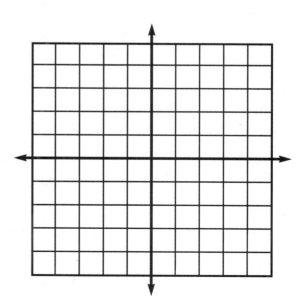

3.

$Y_1 = \sin x$

$Y_2 = \sin x + 1$

$Y_3 = \sin x - 1$

$Y_4 = \sin x + 2$

   a. How are these curves alike? Different?

   b. Graph $y = \sin x + 2$ and $y = 2 + \sin x$. Is there a difference? Why?

   c. Is there a difference between $y = 2 \sin x$ and $y = 2 + \sin x$? Why?

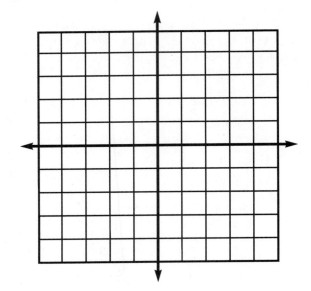

4.

$Y_1 = \sin x$

$Y_2 = \sin 2x$

$Y_3 = \sin \dfrac{1}{2} x$

$Y_4 = \sin 3x$

   a. How are the curves alike? Different?

   b. What effect does changing the coefficient B have on the graph of $y = \sin B$?

   c. Use the calculator to explore $y = \sin Bx$ where $B > 1$ and where $0 < B < 1$. What effect does changing B have on the graph?

   d. Is there a difference between $y = 2 \sin x$ and $y = \sin 2x$? Why?

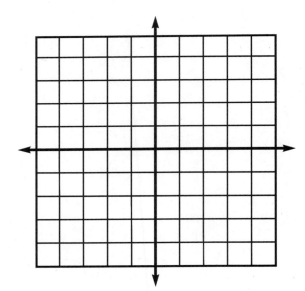

Graphing Power © Dale Seymour Publications

5.

$Y_1 = \sin x$

$Y_2 = \sin (x - 60°)$

$Y_3 = \sin (x + 60°)$

a. What effect does diminishing the argument by 60° have on the graph of $y = \sin x$?

b. What effect does increasing the argument by 60° have on the graph of $y = \sin x$?

c. Describe the effect of C on $y = \sin x$ in $y = \sin (x - C)$.

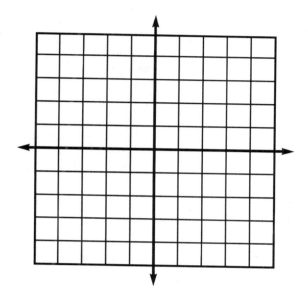

6. Predict how the graph of $y = 3 \sin 2(x - 90°) - 1$ will differ from the graph of $y = \sin x$.

a. Graph $y = 3 \sin 2(x - 90°) - 1$ to test your prediction.

b. Describe the effect of each constant in $y = A \sin B(x - C) + D$.

A

B

C

D

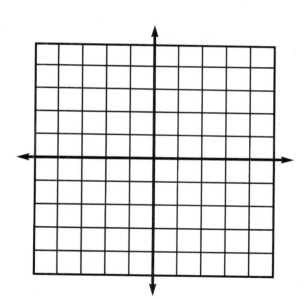

Combinations of Trigonometric Functions

Level   Precalculus, Calculus

Using "Trig Combos" students will

- Use the graphing capabilities of the calculator to explore the addition of trigonometric functions

- Observe the results of raising trigonometric functions to powers

- Make and test conjectures using the graphing calculator

Teaching the Lesson

- To graph these functions, set the calculator to RAD in the $\boxed{\text{MODE}}$ menu and select TRIG in the $\boxed{\text{ZOOM}}$ menu. Change the Y_{scl} to 0.5.

- Deselecting equations will make it easier for students to identify the functions.

- Clear all $\boxed{\text{Y=}}$ entries before starting each new problem set.

- Note: $\sin^2 x$ must be keyed in on the calculator as $(\sin x)^2$.

Trig Combos

1. $Y_1 = \sin x$ $Y_2 = \sin^2 x.$

   $Y_3 = 2\sin x$ $Y_4 = (2\sin x)^2$

   a. Predict what you think the graphs will look like.

   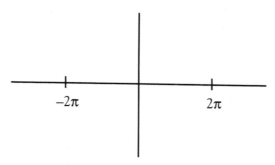

   $\boxed{\text{GRAPH}}$ the functions. Sketch and label the graphs.

   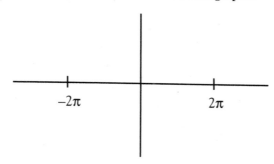

   b. The range of the sin x graph is $-1 \le y \le 1$. What is the range of $y = \sin^2 x$?

   c. The range of $y = 2\sin x$ is $-2 \le y \le 2$ What is the range of $y = (2\sin x)^2$?

2. $Y_1 = \sin x$ $Y_2 = \sin^3 x.$

   a. Predict what you think the graphs will look like.

   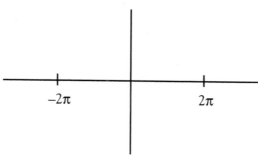

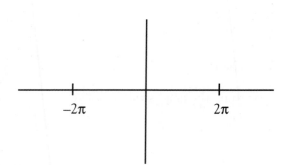

GRAPH both functions. Sketch and label the graphs.

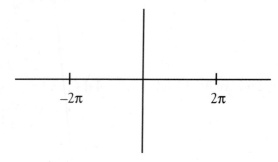

b.   What is the striking difference between the graphs of $y = \sin^2 x$ and $y = \sin^3 x$?

c.   Which other powers of $\sin x$ graphs will resemble the graph of $\sin^3 x$?

3. $Y_1 = \sin x$ $Y_2 = \sin x^2.$

a.   Predict what you think the graphs will look like.

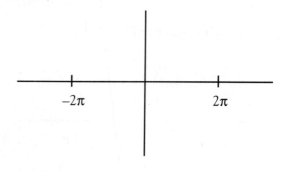

Graphing Power © Dale Seymour Publications

GRAPH both functions. Sketch and label the graphs.

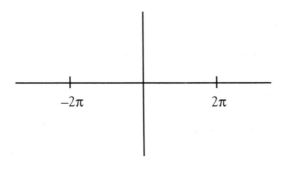

b. What characteristic of the $y = \sin x^2$ curve changes most noticeably as x increases?

4. $Y_1 = \sin x$ $Y_2 = x \sin x.$

a. Predict what you think the graphs will look like.

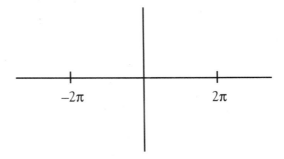

GRAPH both functions. Sketch and label the graphs.

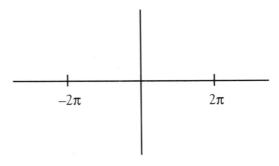

b. What characteristic of the curve changes most noticeably as x increases? Why? ZOOM out for the big picture.

Graphing Power © Dale Seymour Publications

5. Access the ZOOM menu and select TRIG. Go to RANGE and set the Y_{scl} to 1.

$$Y_1 = \sin x \qquad\qquad Y_2 = \frac{1}{x} \qquad\qquad Y_3 = \frac{\sin x}{x}$$

a. GRAPH all three functions. Sketch and label.

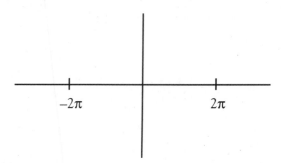

b. What is the apparent value of $\sin x$ at $x = 0$?

c. What is the apparent value of $\dfrac{1}{x}$ at $x = 0$?

d. What is the apparent value of $\dfrac{\sin x}{x}$ at $x = 0$?

e. Explain why the function results in this value as x approaches 0.

6. $Y_1 = \sin x \quad Y_2 = \cos x \qquad Y_3 = Y_1 + Y_2.$

a. GRAPH all three functions. Sketch and label.

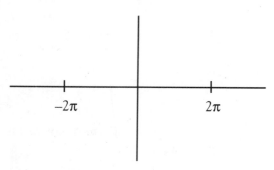

Graphing Power © Dale Seymour Publications

In the $\boxed{Y=}$ menu let $Y_4 = Y_1 - Y_2$. $\boxed{\text{GRAPH}}$ Y_3 and Y_4. Sketch the graphs.

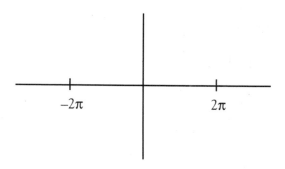

b. Describe the difference between the graphs of $y = \sin x + \cos x$ and $y = \sin x - \cos x$.

7. $Y_1 = \sin x$ $\qquad$ $Y_2 = x$ $\qquad$ $Y_3 = Y_1 + Y_2$

Change the $\boxed{\text{RANGE}}$ values so that $Y_{min} = -9$ and $Y_{max} = 9$.

a. Predict what you think the graph will look like.

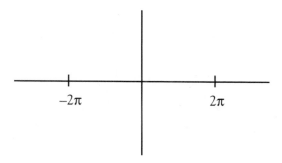

$\boxed{\text{GRAPH}}$ all three functions. Sketch and label.

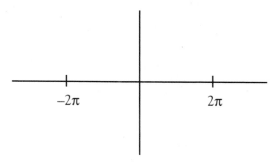

b. Describe the effect of adding $y = x$ to $y = \sin x$.

Trigonometric Identities

Level   Trigonometry, Precalculus, Calculus

Using "Graphical Representation of Identities" students will

- Use the graphing capabilities of the calculator to graph trigonometric functions

- Determine whether or not a set of functions are trigonometric identities

Teaching the Lesson

- To graph these functions, set the calculator to RAD in the $\boxed{\text{MODE}}$ menu and select TRIG in the $\boxed{\text{ZOOM}}$ menu.

- Clear all $\boxed{\text{Y=}}$ entries before starting each new problem set.

- Enter the left side of the equation for Y_1 and the right side for Y_2. If the graphs coincide, then the equation is an identity.

Solutions

1. yes   8. no

2. yes   9. yes

3. no   10. yes

4. yes   11. yes

5. yes   12. yes

6. yes   13. no

7. yes   14. no

Graphical Representation of Identities

Enter the left side of the equation for Y₁ and the right side of the equation for Y₂ and
GRAPH . If the graphs coincide, the equation is an identity. Draw a sketch of each
equation/identity on this sheet.

1. $(\sin x)^2 + (\cos x)^2 = 1$

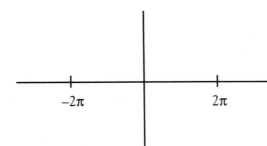

2. $1 + (\tan x)^2 = (\sec x)^2$

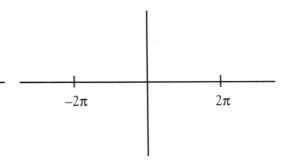

3. $1 + (\cot x)^2 = (\csc x)^2$

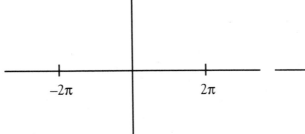

4. $\sin 2x = 2 \sin x \cos x$

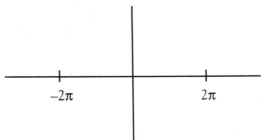

5. $\cos 2x = (\cos x)^2 - (\sin x)^2$

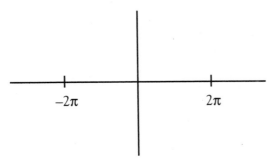

6. $\cos 2x = 2(\cos x)^2 - 1$

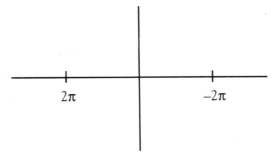

7. $\cos 2x = 1 - 2(\sin x)^2$

8. $\cot x \cdot \sin x = \cos x$

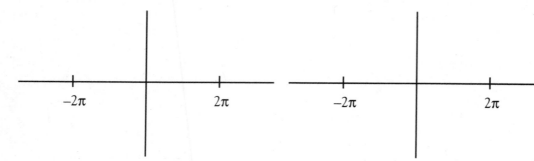

9. $\tan 2x = \dfrac{2 \tan x}{1 - (\tan x)^2}$

10. $\dfrac{\sin x}{\cos x} = \tan x$

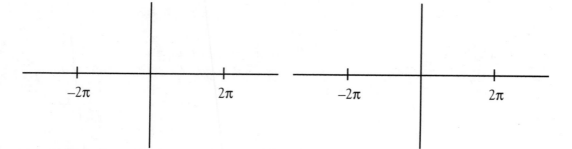

11. $\sin 3x = 3 \sin x$

12. $3 \sin x \cos x = \sin 3x$

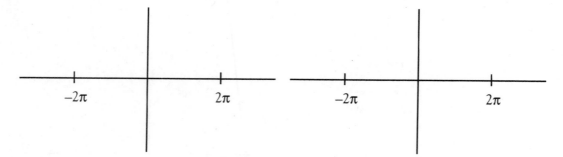

13. $\tan x \cdot \cot x = 1$

14. $\dfrac{\sin 2x}{\sin x} = \cos x$

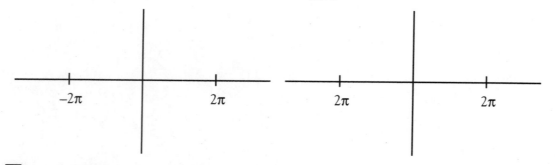

Graphing Power © Dale Seymour Publications

Polar Functions

Level   Analytical Geometry, Precalculus

Using "Polars and Parametrics" students will

- Use the graphing capabilities of the calculator to explore $r = A + B \cos \theta$ and $r = B \cos C\theta$, observing the effect of changing the values of A, B, and C

- Make and test conjectures regarding the relationship of A and B in conjunction with each other in $r = A + B \cos \theta$

- Investigate the affect of C when C is not an integer

Teaching the Lesson

- Select PARAMetric and POLAR in the $\boxed{\text{MODE}}$ menu if using the TI-81.
- You need only select PARAMetric in the $\boxed{\text{MODE}}$ menu if using the TI-82.
- Use the EDIT features to save time.

Solutions

1a.   cardioid
  b.   loop inside cardioid
  c.   ellipse
  d.   loop from bottom
  e.   cardioid
  f.   loop inside
  g.   ellipse

2a.   4 petals, 3 units long, 2 units wide
  b.   mirror image of 2a
  c.   3 petals, 3 units long, 2 nits wide
  d.   mirror image of 2c
  e.   C determines the number of petals.
  f.   mirror image of the positive value
  g.   petal placement in relation to the axis

3g.   C determines the number of petals.
  h.   20, 13
  i.   If C is even, the number of petals is 2C.
        If C is odd, the number of petals is C.

Polars and Parametrics

A polar equation $r = f(\theta)$ can be graphed using the parametric equations

$$x_1 = r \cos \theta$$
$$y_1 = r \sin \theta$$

- Enter the expressions in the [Y=] menu to define the parametric equations in terms of T or θ.

    On the TI-81 for r = 0.5θ, enter $x_1 = 0.5T \cos T$ and $y_1 = 0.5T \sin T$.

    On the TI-82 for r = 0.5θ, enter $r_1 = 0.5\theta \cos \theta$ and $r_2 = 0.5\theta \sin \theta$.

- Press [GRAPH]. It will graph for T or $\theta = 2\pi$. Therefore, go to the [RANGE] menu and change T_{max} to 25. Press [GRAPH] again to see the new graph.

    [GRAPH] each of the following equations. Observe the changes in each graph as you change constants, then answer the questions. Change the range to $T_{max} = 2\pi$, $X_{min} = -5$, $X_{max} = 5$, $Y_{min} = -5$, $Y_{max} = 5$. Use the [ZOOM] menu to select a Square screen.

1. Polar equations of the form $r = a + b \cos \theta$

   a. $r = 2 + 2 \cos \theta$ b. $r = 2 + 3 \cos \theta$
       remember $x_{1T} = (2 + 2 \cos T) \cos T$ and $y_{1T} = (2 + 2 \cos T) \sin T$

   c. $r = 3 + \cos \theta$ d. $r = -2 + 3 \cos \theta$

   Explain the following in reference to the above equations.

   e. If A = B, then

   f. If A < B, then

   g. If A > B, then

Graphing Power © Dale Seymour Publications

2. Polar equations of the form $r = B \cos C\theta$

    a. $r = 3 \cos 2\theta$ b. $r = -3 \cos 2\theta$

    c. $r = 3 \cos 3\theta$ d. $r = -3 \cos 3\theta$

Explain the following in reference to the above equations.

    e. What does C do?

    f. If B changes sign, what happens to the polar graph?

    g. What will change if the function changes from $\cos 2\theta$ to $\sin 2\theta$?

3. Polar equations of the form $r = 3 \cos C\theta$

    a. $r = 3 \cos 3\theta$ b. $r = 3 \cos 4\theta$

    c. $r = 3 \cos 5\theta$ d. $r = 3 \cos 6\theta$

    e. $r = 3 \cos 7\theta$ f. $r = 3 \cos 8\theta$

Explain the following in reference to the above equation.

    g. What does C do?

    h. How many petals will the graph have if C = 10? If C = 13? Explain why.

    i. Predict what any value of C will do. Explain.

    j. How many petals does $r = 3 \sin 3.5\theta$ have? $r = 3 \sin 3.2\theta$? $r = 3 \sin 3.8\theta$? Explain.

Limits

Level   Precalculus, Calculus

Using "In the Neighborhood" students will

- Use the graphing capabilities to explore functions at the point where the limit needs to be determined

- Use substitution as well as the ⸢TRACE⸣ and ⸢ZOOM⸣ menus to explore functions

Teaching the Lesson

- Select Rad in the ⸢MODE⸣ menu before starting this lesson.

Solutions

1a. 0
 b. 3
 c. undefined
 d. undefined
 e. 0
 f. $-\dfrac{1}{2}$
 g. $\dfrac{2}{5}$
 h. 0

4b. at $x = 2$
 c. at $x = -2$
d–g. at $x = 0$
 h. at $x = 1$

5. No, c and d have vertical asymptotes at $x = -2$ and $x = 0$, respectively.

In the Neighborhood

1. Graph the following functions and estimate the limiting value L using the ZOOM and TRACE menus near the value of x.

   a. $\lim\limits_{x \to -2} \quad 2x^2 - 5x + 2$

   b. $\lim\limits_{x \to -2} \quad \dfrac{2x^2 - 5x + 2}{x - 2}$

   c. $\lim\limits_{x \to -2} \quad \dfrac{x + 1}{x + 2}$

   d. $\lim\limits_{x \to 0} \quad \dfrac{x^3 + 9}{6x}$

   e. $\lim\limits_{x \to \infty} \quad \dfrac{1}{x}$

   f. $\lim\limits_{x \to 0} \quad \dfrac{\cos x - 1}{x^2}$

   g. $\lim\limits_{x \to 0} \quad \dfrac{\tan 2x}{5x}$

   h. $\lim\limits_{x \to 1} \quad \dfrac{x^3 - 4x^2 + 5x - 2}{|x - 1|}$

2. Make a table for a sequence of at least six values near x to provide evidence for the limits you estimated in question 1.

   a.
x	$f(x)$

   b.
x	$f(x)$

   c.
x	$f(x)$

   d.
x	$f(x)$

   e.
x	$f(x)$

   f.
x	$f(x)$

g.

x	$f(x)$

h.

x	$f(x)$

3. Were all the functions continuous?

4. If not, list the functions which were not continuous and explain why.

5. Of those that were not continuous, did they all have a limit at the specified x? Explain and illustrate your explanation with an example.

Graphing Power © Dale Seymour Publications

Continuity

Level   **Precalculus, Calculus**

Using "Broken-Up" the student will

- Use the ⎡TRACE⎤ menu to observe where discontinuity exists in a function
- Observe the difference between removable and non-removable (jump and point) discontinuity

Solutions

1. continuous

2. jump

3. removable at $x = 3$

4. jump at $x = 2$, connected elsewhere

5. jump at $x = \frac{\pi}{4}$, π, and at every π thereafter

6. discontinuous at $x = 0$, function not defined at $x = 0$, removable at $(0, 1)$

Broken Up

Graph each function and use the graph to decide where the function is continuous. Where the function is discontinuous, is the discontinuity removable or a jump? If it is removable, give the value of x at which the removable discontinuity occurs.

1. $f(x) = |\, x^4 + x^3 - 2x^2 \,|$

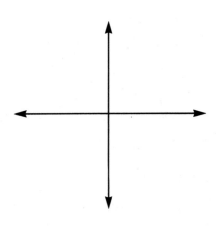

2. $f(x) = \dfrac{1}{x^2 - 9}$

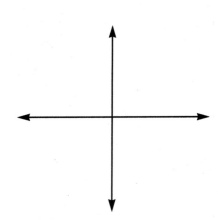

3. $f(x) = \dfrac{x^2 - 6x + 9}{x - 3}$

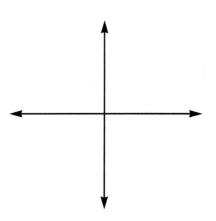

Graphing Power © Dale Seymour Publications

4.

$$f(x) = \begin{cases} 1 & \text{for } x < 2 \\ x + 1 & \text{for } 2 \leq x \leq 5 \\ 31 - x^2 & \text{for } x > 5 \end{cases}$$

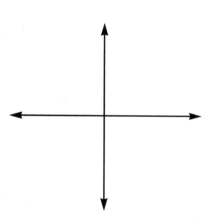

5.

$$f(x) = \begin{cases} \sin x & \text{for } x \leq \dfrac{\pi}{4} \\ \cos x & \text{for } -\dfrac{\pi}{4} < x \leq \pi \\ \cot x & \text{for } x > \pi \end{cases}$$

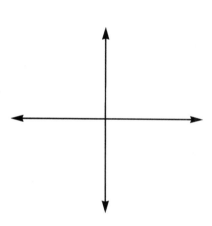

6. $f(x) = \begin{cases} x + 1 & \text{for } x < 0 \\ 1 - x & \text{for } x > 0 \end{cases}$

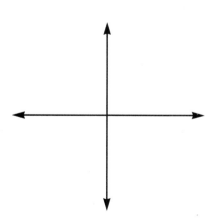

Numerical Derivatives

Level   Calculus

Using "Estimating Derivatives" students will

- Use substitution to approximate the limit approaching from the right and the left

- Gain a visual understanding of the concept of a limit using the graphing calculator

- Compare numerical approximations to the exact values of the derivatives

Solutions

1. 3

2. 0

3. 3

4. (1) $f'(x) = 3x^2$
    (2) $f'(x) = 2x - 2$
    (3) $f'(x) = 3(x - 3)^2$

6. (1) $y = 3x - 2$
    (2) $y = 0$
    (3) $y = 3x - 7$

8. Slope of the tangent line is the slope of the curve at the point of tangency.

Estimating Derivatives

Use the calculator to numerically estimate the value of D at the given value of x using values of h such that $-0.5 \le h \le .05$.

1. $f(x) = x^3$ at $x = 1$

$$D = \lim_{h \to 0} \frac{(x + h)^3 - x^3}{h}$$

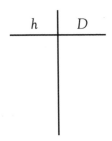

h	D

$D =$

2. $f(x) = x^2 - 2x$ at $x = 1$

$$D = \lim_{h \to 0} \frac{[(x + h)^2 - 2(x + h)] - [x^2 - 2x]}{h}$$

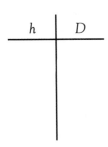

h	D

$D =$

Graphing Power © Dale Seymour Publications

3. $f(x) = (x - 3)^3$ at $x = 2$

$D =$

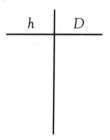

$D =$

4. Evaluate the limits for questions 1–3 algebraically to find the value of D.

5. In your own words, define a derivative of a function $f(x)$ at x_1 and draw an illustration that demonstrates why the limit as $h \rightarrow 0$ is necessary in the definition.

6. Find the equation of the line tangent to the function at the given x for each of the functions in questions 1–3.

7. Using the graphing calculator, sketch each of the functions in questions 1–3 and the tangent lines you obtained in question 6.

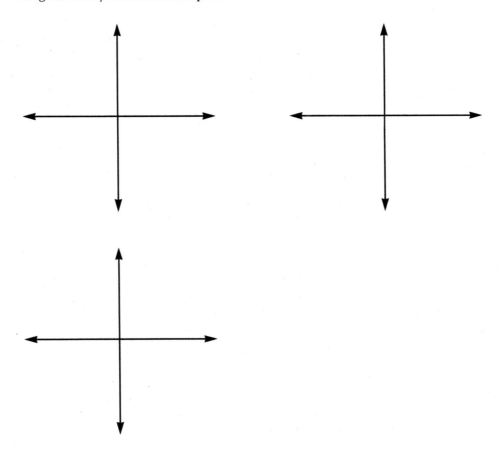

8. What do you notice about the slope of the tangent line and the curvature of the graph?

Infinite Series and Convergence

Level   Algebra II, Geometry, Precalculus

Using "Limits" students will

- Use graphing technology to observe convergence, divergence, and sum of a convergent infinite series

- Use the $\boxed{\text{TABLE}}$ menu to determine the sum of a series

Teaching the Lesson

- The TI-82 is required for this lesson.

- Select Seq in the $\boxed{\text{MODE}}$ menu.

- Clear U_n and V_n equations in the $\boxed{\text{Y=}}$ menu.

- Knowledge of the $\boxed{\text{WINDOW}}$ settings for sequences is imperative to the lesson.

U_{n-1}, V_{n-1}, and n are 2nd function keys located above the $\boxed{7}$, $\boxed{8}$, and $\boxed{9}$.

- Set the TblSet to: Set $\boxed{\text{WINDOW}}$ to:

    TblMin = 11 U_nStart = 1

    ΔTbl = 1 V_nStart = 1

    $\lim$ lpnt: Auto   nStart = 0

    Depend: Auto   nMin = 1

                    nMax = 25

                    X_{min} = 1

                    X_{max} = 1.5

- Let U_n = Series

- Let V_n = Series + V_{n-1}

Solutions

1. converges, $2 + \sqrt{2}$
2. diverges
3. converges, $\dfrac{5}{6}$
4. converges, $\dfrac{2}{9}$
5. converges, $\dfrac{3}{2}$

6. diverges
7. diverges
8. converges, 2

Limits

Determine graphically whether each series converges or diverges. Use the Table function to find the sum of the series if it converges.

1. $\displaystyle\sum_{n=0}^{\infty} \left(\frac{1}{\sqrt{2}}\right)^n$

2. $\displaystyle\sum_{n=1}^{\infty} (\sqrt{2})^n$

3. $\displaystyle\sum_{n=0}^{\infty} \frac{\cos n\pi}{5^n}$

4. $\displaystyle\sum_{n=1}^{\infty} \frac{2}{10^n}$

5. $\displaystyle\sum_{n=0}^{\infty} \frac{2^n - 1}{3^n}$

6. $\displaystyle\sum_{n=1}^{\infty} \left(1 - \frac{1}{n}\right)^n$

7. $\displaystyle\sum_{n=1}^{\infty} \ln\frac{1}{n}$

8. $\displaystyle\sum_{n=1}^{\infty} \frac{1}{2^n}$

Integration

Level   Precalculus

Using "Definite Integrals" students will

- Use graphing technology to analyze definite integrals
- Use the integration function in the CALC and MATH menus to analyze integrals
- Calculate definite integrals for several types of integration problems

Teaching the Lesson

- The TI-82 is required for this lesson.
- Clear all equations in the Y= menu.
- Reset the WINDOW using ZStandard in the ZOOM menu.
- In the MATH menu, the parameters of the integration functions are fnInt(function, variable, begin, end).
- To graph the integral, access the Y= menu and enter fnInt(function, variable, begin, end) for Y_1.
- The CALC menu will give the approximation and solution based on what is entered into the Y= menu.

Solutions

1. 0.3333

2. This is the exact value.

3. 10.30895

4. You must move the arrow away from $x = 0$, because the function is asymptotic at that value.

6. The integral must be broken into three integrals because of the negative area.

7. fnInt(cos x, x, 0, $\pi/2$,) – fnInt(cos x, x, 0, π) + fnInt(cos x, x, 0, $3\pi/2$)

8. You must subtract the areas of the smaller curve from the area of the larger curve.

9. 7.3333, fnInt($6 - x^2$, x, 0, 2) – fnInt(x, x, 0, 2)

10. Lower limit is 0, and upper limit is 3.19 because of the accuracy of the trace. 12.770883 is the answer. By breaking up the function into $\pi/4$ths, the answer is 12.566376.

11. 3.140300156
    3.140594899
    3.140592651
    3.140592652
    3.140592654 for the last three

12. The last three are the same because the algorithm that solves the integral has already found the most accurate answer it can.

13. The limit beyond which change is very insignificant.

Definite Integrals

Sketch a graph of the definite integrals and find their solutions.

1. $\displaystyle\int_{0}^{1} (x^2 - 2x + 1)\, dx$

2. Does this approximation come close to the actual answer?

3. $\displaystyle\int_{0}^{3} \frac{1}{x}\, dx$

4. What was necessary to calculate this answer? Why? Is your answer a good one?

5. $\displaystyle\int_{0}^{\frac{3\pi}{2}} \cos x\, dx$

6. What did you do to calculate question 5? Why?

7. What did you enter into your calculator to do question 5?

8. How is the area between curves found?

9. Graph the functions $y = 6 - x^2$ and $y = x$.

   a.   Find the area between the two functions from $x = 0$ to $x = 2$.

   b.   What did you enter into your calculator?

Graphing Power © Dale Seymour Publications

10. $\int_{0}^{\pi} (4 + 2 \sin 2x \cos 4x)\, dx$

(Use the CALC menu function first.)

The calculator allows us to find the definite integral of a problem that may appear unsolvable. However the accuracy of the calculator decreases with an erratic function like this. To counter this, break up the integral into smaller intervals. Use the fnInt function to find a more accurate answer.

The program used by the TI 82 is the Gauss-Kronrod method of numerical integration. The tolerance (accuracy) of this is 1×10^{-5}. The tolerance can be specified in the fnInt function like this:

$$\text{fnInt } (f(x), x, a, b, \text{tolerance}).$$

11. Try adjusting the tolerance on fnInt $\frac{1}{(1+x^2)}$, x, -2000, 2000, tol). Record your answers.

Tolerance	Answer
1×10^{-1}	
1×10^{-2}	
1×10^{-3}	
1×10^{-4}	
1×10^{-7}	
1×10^{-10}	
1×10^{-20}	

12. How do the last three compare? Why?

13. The domain of [−2000, 2000] was not chosen arbitrarily. What domain does it represent?

Graphing Power © Dale Seymour Publications

TOPIC 9

Programming for the TI-81 and TI-82

Sample Programs

Level   Algebra, Geometry, Advanced Math

Using these sample programs students can

- See how everyday, time-consuming mathematical operations can be programmed

- Analyze the purpose of each step in a program

- Test the programs

- Evaluate programs for efficient use of available memory

- Modify the programs

- Use the programs as examples, as they translate other mathematical jobs into machine commands

Teaching with the Sample Programs

- The language used by the graphing calculator has some limitations, but the ease of working with commands makes it possible to program using minimum time on computer commands and maximum time on the mathematical process.

- Encourage students to modify and test modifications for any program as they experiment with programming. They can use the sample programs as models. For example, they can apply formula programs to new formulas.

- Discuss with your students the balance between efficient use of memory and understandability of the program.

- If students use their own machine, they can store programs. Students who share calculators may find some programs already entered and learn from other students' work.

- Some of the sample programs are given names. You can change these names as needed to work with other programs saved on your graphing calculator.

- For some examples, the programs for the TI-81 and TI-82 are similar; for others, they do different things. Subroutines on the TI-81 are referenced by number, Prgm8. Subroutines on the TI-82 are referenced by name.

- There are some differences between programming the TI-81 and TI-82. On the TI-82

    END has been replaced with STOP because the END statements have a new purpose.

    It is necessary to rewrite comments on Disp statements to 16 characters or less.

    When you are making changes to the range values, press WINDOW .

    When you press STO▶ you must press ALPHA to get into alpha mode.

    Programs can be directly transferred from one TI-82 to another.

Root Estimates

These programs will compute the square root of a number using the divide and average method. The student estimates the first square root.

TI-81

```
Prgm:SQROOT
:Int(Rand*100)→X
:Disp Ans
:Disp "GUESS THE SQUARE ROOT"
:Input A
:Lbl 1
:ClrHome
:X/A→B
:Disp B
:Disp A
:Pause
:100000→J
:If Int JB=Int JA
:Goto W
:(B+A)/2→A
:Goto 1
:Disp A
:Lbl W
:ClrHome
:Disp B
:Disp A
:Disp "SQROOT OF"
:Disp X
:End
```

TI-82

```
PROGRAM: SQROOT
:Lbl S
:int (rand*100)→X
:If X=0
:Goto S
:Disp "GUESS THE "
:Disp "SQ ROOT OF", X
:Input A
:Lbl 1
:ClrHome
:X/A→B
:Disp B
:Disp A
:Disp "PRESS ENTER"
:Disp "TO CONTINUE"
:Pause
:100000000→J
:If int JB = int JA
:Goto W
:(B+A)/2→A
:Goto 1
:Disp A
:Lbl W
:ClrHome
:Disp B
:Disp A
:Disp "SQ ROOT OF"
:Disp X
:Disp "IS", √X
:Stop
```

Star

This program explores a star fractal pattern. Notice the display features on the TI-82 program.

TI-81

```
:π→G
:-.4→Xmin
:1.5→Xmax
:-.8→Ymin
:.7→Ymax
:5→P
:4→V
:144→A
:.35→R
:AG/180→A
:0→K
:0→L
:0→X
:0→Y
:0→N
:Disp N
:Lbl 1
:N→M
:N*A→B
:0→F
:Lbl 2
:If M/V≠Int M/V
:Goto 3
:If F≥P-1
:Goto 3
:F+1→F
:Int M/V→M
:Goto 2
:Lbl 3
:X+R^(P-F-1)cosB->X
:Y+R^(P-F-1)sinB->Y
:Line(K,L,X,Y)
:X→K
:Y→L
:N+1→N
:If N≤320
:Goto 1
:Pause
:ClrDraw
```

TI-82

```
PROGRAM: STAR
:CoordOFF
:AxesOff
:π→G
:-.4→Xmin
:1.5→Xmax
:-.8→Ymin
:.7→Ymax
:5→P
:4→V
:144→A
:.35→R
:AG/180→A
:0→K
:0→L
:0→X
:0→Y
:0→N
:Disp N
:Lbl 1
:N→M
:N*A→B
:0→F
:Lbl 2
If M/V≠int M/V
:Goto 3
:If F≥ P- 1
:Goto 3
:F+1→F
:int M/V→M
:Goto 2
:Lbl 3
:X+R^(P-F-1)cosB→X
:Y+R^(P-F-1)sin B→Y
:Line(K,L,X,Y)
:X→K
:Y→L
:N+1→N
:If N≤320
:Goto 1
:Pause
:ClrDraw
```

Graphing Power © Dale Seymour Publications

Area Under a Curve

These programs are used to find the area under a curve from a left boundary on the *x*-axis to a right boundary on the *x*-axis. The function used must be under Y_1 in Y= menu.

TI-81

```
:Lbl 1
:0→A
:Disp "LEFT BOUNDARY"
:Input L
:Disp "RIGHT BOUNDARY"
:Input R
:Disp "X INCREMENT"
:Disp "BEST INCREMENT IS .1"
:Input I
:Lbl 2
:If L≥R
:Goto 3
:L→X
:Y₁→M
```
(Access Y_1 by pressing 2nd Y-VARS 1 ENTER)
```
:X+ I →X
:Y₁→N
:A+.5I(M+N)→A
```
    (Trapezoidal method)
```
:X→L
:Disp A
:Goto 2
:Lbl 3
:Disp "AREA"
:Disp A
:Pause
:ClrHome
:Goto 1
```

TI-82

```
PROGRAM:ACURVE
:Lbl 1
:0→A
:Disp "LEFT BOUNDARY"
:Input L
:Disp "RIGHT BOUNDARY"
:Input R
:Disp "X INCREMENT"
:Input "BEST IS .1", I
:Lbl 2
:If L≥R
:Goto 3
:L→X
:Y₁→M
:X+I→X
:Y₁→N
:A+.5I(M+N)→A
:X→L
:Goto 2
:Lbl 3
:Disp "AREA",A
:Pause
:ClrHome
:Goto 1
```

Solve Quadratic Equations

This program will solve for x in a quadratic equation expressed in standard form
($y = ax^2 + bx + c$, $a \neq 0$) using the quadratic formula. The function is then drawn for y equals
the quadratic expression by pressing ENTER after the values of x have been found.

Set Range values or build extra program lines into the program to set the range for you.
$X_{min}=-9.4$, $X_{max}=9.6$, $X_{scl}=1$, $Y_{min}=-6.2$, $Y_{max}=6.4$, $Y_{scl}=1$, $X_{res}=1$

TI-81

```
:ClrDraw
:ClrHome
:Disp "A"
:Input A
:Disp "B"
:Input B
:Disp "C"
:Input C
:(B²-4AC)→D
:If D<0
:Goto 1
:If D=0
:Goto 2
:(-B+√D)/2A→X
:(-B-√D)/2A→Z
:Disp "X1"
:Disp X
:Disp "X2"
:Disp Z
:Pause
:DrawF AX²+BX+C
    ( 2nd  DRAW  6  ENTER )
:End
:Lbl 1
:Disp "COMPLEX"
:Pause
:DrawF AX2+BX+C
:End
:Lbl 2
:Disp "ONE FACTOR"
:-B/2A->X
:Disp "X1="
:Disp X
:Pause
:DrawF AX²+BX+C
:End
```

TI-82

```
PROGRAM:QUAD
:CoordOn
:RectGC
:AxesOn
:-9.4→Xmin
:9.6→Xmax
:1→Xscl
:-6.2→Ymin
:6.4→Ymax
:1→Yscl
:ClrDraw
:ClrHome
:Input "A", A
:Input "B", B
:Input "C", C
:(B²-4AC)→D
:If D≥0
:Then
:(-B+√D)/2A→Y
:(-B-√D)/2A→Z
:Disp "X1", Y
:Disp "X2", Z
:Else
:Disp "COMPLEX"
:End
:Disp "PRESS ENTER"
:Disp "TO GRAPH"
:Pause
:DrawF AX²+BX+C
```

Graphing Power © Dale Seymour Publications

General Conics

These programs will graph all conics using the general polynomial form
$Ax^2 + Bxy + Cy^2 + Dx + Ey + F = 0$.

TI-81

```
:-9.4→Xmin
:9.6→Xmax
:-6.4→Ymin
:6.2→Ymax
:GridOn
     ([MODE] [>] 8 [ENTER])
:ClrHome
:ClrDraw
:Disp "A"
:Input A
:Disp "B"
:Input B
:Disp "C"
:Input C
:Disp "D"
:Input D
:Disp "E"
:Input E
:Disp "F"
:Input F
:If C=0
:Goto 2
:DrawF((-BX-E)+√((BX+E)²-4C(AX²+DX+F)
     ))/2C
:DrawF((-BX-E)-√((BX+E)²-4C(AX²+DX+F)
     ))/2C
:End
:Lbl 2
:DrawF(AX²+DX+F)/(-BX-E)
:End
```

TI-82

```
:PROGRAM:CONIC
:ClrDraw
:DispGraph
:Text(0,0, "PROGRAM WILL GRAPH")
:Text(10,0, "CONICS IN THE FORM")
:Text(20,0, "AX²+BXY+CY²+DX+EY+F=0")
:Text(30,0, "PRESS ENTER")
:Pause
:GridOn
:AxesOn
:ClrDraw
:-9.4→Xmin
:9.6→Xmax
:-6.4→Ymin
:6.2→Ymax
:Input  "A", A
:Input  "B", B
:Input  "C", C
:Input  "D", D
:Input  "E", E
:Input  "F", F
:If C≠0
:Then
:DrawF
     ((-BX-E)+√((BX+E)²-4C(AX²+DX+F)))/
     2C
:DrawF
     ((-BX-E)-√((BX+E)²-4C(AX²+DX+F)))/
     2C
:Else
:DrawF(AX²+DX+F)/(-BX-E)
:End
```

Net Pay

These programs will compute net pay with time-and-a-half for overtime, 8% for FICA and 12% for withholding tax.

TI-81

```
:Fix 2
:ClrHome
:Disp "PAY PER HOUR"
:Input P
:Disp "HOURS WORKED"
:Input H
:If H>40
:Goto 3
:PH→G
:Lbl 4
:Disp "GROSS PAY"
:Disp G
:Pause
:.08G→S
:.12G→I
:G-S-I→N
:Disp "FICA TAX"
:Disp S
:Disp "FED. WITH. TAX"
:Disp I
:Disp "NET INCOME"
:Disp N
:End
:Lbl 3
:(H-40)*1.5*P→E
:40P+E→G
:Goto 4
```

TI-82

```
PROGRAM:NETPAY
:Fix 2
:ClrHome
:Input "PAY PER HOUR", P
:Input "HOURS WORKED", H
:If H > 40
:Then
:1.5P(H-40)+40P→G
:Else
:PH→G
:End
:ClrHome
:Output (1,1, "GROSS PAY")
:Output (1,10,G)
:.08G→S
:.12G→I
:G-S-I→N
:Output (2,1, "FICA TAX")
:Output (2,10,S)
:Output (3,1, "FED. TAX")
:Output (3,10,I)
:Output (4,1, "NET INCOME")
:Output (5,10,N)
```

Graphing Power © Dale Seymour Publications

Distance Formula

These programs use the distance formula

$$\sqrt{(x_1 - x_2)^2 + (y_1 - y_2)^2}$$

to find the distance between the two points (x_1, y_1) and (x_2, y_2).

TI-81

```
:ClrDraw
:Disp "X1="
:Input Q
:Disp "Y1="
:Input R
:Disp "X2="
:Input S
:Disp "Y2="
:Input T
:Line(Q,R,S,T)
:Pause
:√((Q-S)²+(R-T)²)→D
:Disp "DIST ="
:Disp D
```

TI-82

```
PROGRAM:DISTANCE
:ClrDraw
:Input "X1=", Q
:Input "Y1=", R
:Input "X2=", S
:Input "Y2=", T
:Line (Q,R,S,T)
:√((Q-S)²+(R-T)²→D
:Text (5,50, "DIST=", D)
```

Slope Formula

These programs find the slope of a line containing any two points, (x_1, y_1) and (x_2, y_2), using the slope formula, slope = $\frac{(x_1 + x_2)}{2}$. The program for the TI-82 also displays the equation of the line in the slope-intercept form.

TI-81

```
:ClrDraw
:Disp "X1="
:Input Q
:Disp "Y1="
:Input R
:Disp "X2="
:Input S
:Disp "Y2="
:Input T
:If (Q-S)=0
:Goto 2
:(R-T)/(Q-S)→A
:Line(Q,R,S,T)
:Pause
:Disp "SLOPE ="
:Disp A
:End
:Lbl 2
:Disp "SLOPE UNDEF"
```

TI-82

```
PROGRAM:SLOPE
:ClrDraw
:Input "X1=", Q
:Input "Y1=", R
:Input "X2=", S
:Input "Y2=", T
:If (Q-S)≠0
:Then
:(R-T)/(Q-S)→A
:R-AQ→B
:DrawF AX+B
:Text (2,5, "EQUATION IS")
:Text (8,5, "Y=", A, "X+", B)
:Else
:ClrHome
:Text (1,1, "SLOPE UNDEFINED")
:Text (45,1, "EQUATION OF LINE")
:Text (55,1, "X=", R-AQ)
:Vertical B
```

Graphing Power © Dale Seymour Publications

Midpoint Formula

These programs find the coordinates of the midpoint of a segment given the two endpoints, (x_1, y_1) and (x_2, y_2), using the midpoint formula

$$(\frac{(x_1 + x_2)}{2}, \frac{(y_1 + y_2)}{2})$$

TI-81

```
:ClrDraw
:Disp "X1="
:Input Q
:Disp "Y1="
:Input R
:Disp "X2="
:Input S
:Disp "Y2="
:Input T
:(Q+S)/2→M
:(R+T)/2→N
:Line(Q,R,S,T)
:Pause
:PT-Chg(M,N)
:Disp "MIDPT (M,N) ="
:Disp M
:Disp N
```

TI-82

```
PROGRAM:MIDPOINT
:ClrDraw
:ClrHome
:Input "X1=", J
:Input "Y1=", K
:Input "X2=", L
:Input "Y2=", M
:Line(J,K,L,M)
:(J+K)/2→P
:(L+M)/2→Q
:Text(1,1, "MIDPOINT IS (", M, ",",
    N, ")")
```

Distance, Midpoint, and Slope Formulas

These programs combine the computation of the three formulas into one program and calculate and display the y-intercept of the line that contains the points (x_1, y_1) and (x_2, y_2). In the TI-82 program, you can change the FullScreen to Split and watch the display while you enter data.

TI-81

```
:ClrDraw
:Disp "X1="
:Input Q
:Disp "Y1="
:Input R
:Disp "X2"
:Input S
:Disp "Y2="
:Input T
:Line(Q,R,S,T)
:Pause
:√((Q-S)²+(R-T)²)→D
:Disp "DIST ="
:Disp D
:(Q+S)/2→M
:(R+T)/2→N
:Disp "MIDPT (M,N) ="
:Disp M
:Disp N
:If (Q-S)=0
:Goto 2
:(R-T)/(Q-S)→A
:Disp "SLOPE ="
:Disp A
:R-A*Q→B
:Disp "Y INT="
:Disp B
:End
:Lbl 2
:Disp "SLOPE UNDEF"
:Disp "Y INT-NONE"
```

TI-82

```
PROGRAM:COMBDMS
:Fix 3
:ClrDraw
:FullScreen
:Input "X1=", Q
:Input "Y1=", R
:Input "X2=", S
:Input "Y2=", T
:Line (Q,R,S,T)
:√((Q-S)²+(R-T)²)→D
:Text (5,1, "DIST=", D)
:(Q+S)/2→F
:(R+T)/2→G
:Text (1,1, "MIDPOINT IS (", F, ",",
    G, ")")
:For(D,0,200)
:Pt-Change(F,G)
:End
:If Q-S=0
:Then
:Text(10,1, "SLOPE UNDEFINED")
:Text(24,1, "EQUATION X=", Q)
:Vertical Q
:Else
:(R-T)/(Q-S)→M
:R-MQ→B
:DrawF MX+B
:Text(24,1, "EQUATION Y=", M, "X+",
    B)
:Text(15,1, "Y-INT=", B)
```

Graphing Power © Dale Seymour Publications

Inscribe

This program for the TI-82 utilizes the midpoint formula for displaying triangles. Removing the pause lines will speed up the display of the pattern.

TI-82

```
PROGRAM INSCRIBE
:FullScreen
:-9.4→Xmin
:9.6→Xmax
:-6.4→Ymin
:6.2→Ymax
:ClrDraw
:ClrHome
:Disp "ENTER 3 POINTS"
:Input "X1=", Q
:Input "Y1=", R
:Input "X2=", S
:Input "Y2=", T
:Input "X3=", U
:Input "Y3=", V
:Lbl 2
:Line (Q,R,S,T)
:Pause
:Line (S,T,U,V)
:Pause
:Line (U,V,Q,R)
:Pause
:(Q+S)/2→J
:(R+T)/2→K
:(S+U)/2→L
:(T+V)/2→M
:(U+Q)/2→N
:(V+R)/2→O
:J→Q
:K→R
:L→S
:M→T
:N→U
:O→V
:Goto 2
```

Parabola

These programs graphically display the construction of a parabola using the focus/directrix definition and the equation $y - A(x - H)^2 + K$. Best results will be obtained if the following limits are observed:

$$-1 < A < 1, A \neq 0 \qquad -6 < H < 6$$

TI-81

```
:ClrDraw
:ClrHome
:1→Xscl
:1→Yscl
:1→Xres
:Disp "WHAT IS A FOR YOUR PARABOLA?"
:Input A
:Disp "WHAT IS H FOR YOUR PARABOLA?"
:Input H
:Disp "WHAT IS K FOR YOUR PARABOLA?"
:Input K
:If H≥0
:H-6→Xmin
:If H≥0
:2*(H+6)→Xmax
:If H<0
:2*(H-6)→Xmin
:If H<0
:(H+6)→Xmax
:If abs (Xmax)>abs (Xmin)
:Xmax→Q
:If abs (Xmin)>abs (Xmax)
:-1*(Xmin)→Q
:If A<0
:(-2/3)*Q→Ymin
:If A<0
:(K-(1/4A)+1)→Ymax
:If A>0
:(2/3)*Q→Ymax
:If A>0
:(K-(1/4A)-1)→Ymin
:(H-5)→W
:Lbl 1
:ClrDraw
:ClrHome
:W→X
```

TI-82

```
PROGRAM:PARABOLA
:ClrDraw
:ClrHome
:1→Xscl
:1→Yscl
:Disp "PARABOLA VALUES"
:Disp "LIMITS"
:Disp "-1<A<1, A≠0"
:Disp "-6<H<6"
:Input "A=", A
:Input "H=", H
:Input "K=", K
:If H<0
:Then
:2(H-6)→Xmin
:(H+6)→Xmax
:Else
:2(H+6)→Xmax
:H-6→Xmin
:If abs Xmax>abs Xmin
:Xmax→Q
:If abs Xmin>abs Xmax
:-1Xmin→Q
:If A<0
:Then
:(-2/3)Q→Ymin
:K-(1/4A)+1→Ymax
:Else
:(2/3)Q→Ymax
:K-(1/4A)-1→Ymin
:End
:H-5→W
:Lbl 1
:ClrHome
:ClrDraw
:W→X
```

Graphing Power © Dale Seymour Publications

TI-81 (cont.)

```
:Line(Xmin,(K-(1/4A)),Xmax,(K-(1/4A)))
:PT-On(H,(K+(1/4A)))
:Line(H,(K+(1/4A)),X,(A*(X-H)^2+K))
:Line(X,(K-(1/4A)),X,(A*(X-H)^2+K))
:X+.3→W
:Lbl 2
:PT-On(X,(A*(X-H)^2+K))
:X-.3→X
:If X≥(H-5)
:Goto 3
:If W≤(H+5)
:Goto 1
:End
:Lbl 3
:15→J
:Lbl 4
:J-1→J
:If J>1
:Goto 4
:Goto 2
```

TI-82 (cont.)

```
:Line(Xmin, (K-(1/4A)), Xmax,
    (K-(1/4A)))
:Pt-On(H, (K+(1/4A)))
:Line(H, (K+(1/4A)), X, (A(X-H)²+K))
:Line(X,(K-(1/4A)), X, (A(X-H)²+K))
:X+.3→W
:Lbl 2
:Pt-On(X, (A(X-H)²+K))
:X-.3→X
:If X≥H-5
:Goto 3
:If W≤H+5
:Goto 1
:Stop
:Lbl 3
:For(J,15,1,-1)
:Text(1,1, "   ")
:End
:Goto 2
```

What's My Angle?

Given a point on the terminal vector, this program computes and displays the angle formed with the zero vector. Angles formed that coincide with the x- or y-axis will not be displayed. See next program.

TI-81

```
:ClrDraw
:Disp "GIVEN LINE FROM (0,0) TO (X1,Y1) WHAT ANGLE WILL BE FORMED WITH THE X
    AXIS?"
:Disp "X1"
:Input X
:Disp "Y1"
:Input Y
:Line (X,Y,0,0)
:Pause
:If X=0
:Goto 3
:If Y=0
:Goto 4
```

TI-81 (cont.)

```
:tan-1 (Y/X) →Q
:Disp "ANGLE IS"
:If X<0
:Q + 180 →Q
:If Q <0
:Q + 360 →Q
:Disp Q
:End
:Lbl 3
:If Y>0
:Disp "ANGLE = 90"
:If Y<0
:Disp "ANGLE = 270"
:If Y=0
:Disp "NO ANGLE"
:End
:Lbl 4
:If X>0
:Disp "ANGLE = 0"
:If X <0
:Disp "ANGLE = 180"
```

What's My Angle II

TI-81

```
:Lbl 1
:Deg
:ClrDraw
:Disp "SHOW ME AN ANGLE = TO "
:Input Q
:Line (10cos Q,10sin Q,0,0)
:Pause
:Disp "1 TO CONTINUE"
:Disp "2 TO STOP"
:Input A
:If A = 2
:End
:Goto 1
```

Graphing Power © Dale Seymour Publications

Equations 'R Us

Solves equation in standard form given A, B, and C.

TI-81

```
:0 → K
:Disp "SOLVE AX + B = C A,B,C ARE"
:I Part (Rand*9) +1 →A
:I Part (Rand +.5) →Z
:If Z = 0
:A*-1 →A
:I Part (Rand * 9) +1 →B
:I Part (Rand + .5) →Z
:If Z=0
:B*-1 →B
:I Part (Rand * 9) +1 →C
:I Part (Rand +.5) →Z
:If Z = 0
:C*-1 →C
:Disp A
:Disp B
:Disp C
:Disp "WHAT IS X?"
:(C-B)/A →X
:Lbl 2
:Input G
:If G ≠ X
:GoTo 5
:If G = X
:GoTo 4
:Lbl 4
:Disp "GOOD JOB"
:End
:Lbl 5
:K+1 →K
:If K = 2
:GoTo 8
:Disp "TRY AGAIN"
:GoTo 2
:Lbl 8
:Disp "THE ANSWER IS  "
:Disp X
```

TI-82

```
PROGRAM:EQUATION
:ClrDraw
:ClrHome
:AxesOff
:0→K
:iPart (rand*9)+1→A
:iPart (rand+.5)→Z
:If Z=0
:-A→A
:iPart (rand*9)+1→B
:iPart (rand+.5)→Z
:If Z=0
:-B→B
:iPart (rand*9)+1→C
:iPArt(rand +.5)→Z
:If Z=0
:-C→C
:Lbl 1
:Text (1,1, "SOLVE  ", A, "X+  ", B,
   = ", C)
:Text (10,1,"PRESS ENTER WHEN YOU")
:Text (17,1, "HAVE THE ROOT")
:Pause
:Input X
:If X=(C-B)/A
:Then
:Text (40,1, "GOOD JOB ", X, "IS THE
   ROOT ")
:Else
:Text(40,1, "PRESS ENTER TO TRY
   AGAIN")
:Pause
:ClrDraw
:Goto 1
```

Pythagorean Theorem

Calculates the hypotenuse of a right triangle. The TI-82 program adds graphic display to the computation of hypotenuse by showing the right triangle being measured.

TI-81

```
:ClrHome
:Disp "SIDE A"
:Input A
:Disp " "
:Disp "SIDE B"
:Input B
:A²+B²→C
:√C→C
:Disp " "
:Disp "SIDE C"
:Disp C
```

TI-82

```
PROGRAM:PYTHY
:FnOff
:AxesOff
:-1→Xmin
:-1→Ymin
:2→Xscl
:2→Yscl
:ClrDraw
:ClrHome
:Split
:Input "SIDE A", A
:Line(0,A,0,0)
:Input "SIDE B", B
:Line(B,0,0,0)
:√(A²+B²)→C
:Line(B,0,0,A)
:Text(1,1, "HYPOTENUSE IS",C)
:Shade(0,(-A/B)X+A,1,0,B)
```

Graphing Power © Dale Seymour Publications

Triangle Measures

Given SSS or SAS, these programs find missing angles and side lengths using the law of cosines.

TI–81

```
:ClrHome
:Disp "1. SSS"
:Disp "2. SAS"
:Disp "Choose 1 or 2"
:Input K
:If K=1
:Goto 1
:If K=2
:Goto 2
:Lbl 1
:Disp "ENTER SIDES A,B,C"
:Input A
:Input B
:Input C
:(cos-1((A²+B²-C²)/2AB))*360/2π→F
:(cos-1((B²+C²-A²)/2BC))*360/2π→D
:180-F-D→E
:Disp "ANGLE A"
:Disp D
:Disp "ANGLE B"
:Disp E
:Disp "ANGLE C"
:Disp F
:End
:Lbl 2
:Disp "ENTER SIDES A, B"
:Input A
:Input B
:Disp "ENTER ANGLE C IN DEGREES"
:Input Q
:Q*2π/360→F
:√(A²+B²-2ABCos F)→C
:(cos⁻¹((B²+C²-A²)/2BC))*360/2π→D
:(cos⁻¹((A²+C²-B²)/2AC))*360/2π→E
:Disp "SIDE C"
:Disp C
:Disp "ANGLE A"
:Disp D
```

TI-82

```
PROGRAM:TRIANGLE
:Lbl 1
:FullScreen
:ClrHome
:Menu("METHOD", "SSS", Z, "SAS", Y)
:Lbl Z
:Disp "ENTER SIDES A, B, C"
:Input A
:Input B
:Input C
:Radian
:(cos⁻¹((A²+B²-C²)/2AB))*360/2π→F
:(cos⁻¹((B²+C²-A²)/2BC))*360/2π→D
:180-F-D→E
:If F≥180 or D≥180 or E≥ 180
:Then
:Disp "NO TRIANGLE IS"
:Disp "FORMED WITH SIDES"
:Goto 1
:Else
:Disp ANGLE A", D
:Disp ANGLE B", E
:Disp ANGLE C", F
:Lbl Y
:Disp "ENTER EACH"
:Disp "KNOWN SIDE A,B"
:Input A
:Input B
:Lbl 2
:Disp "ENTER INCLUDED"
:Disp "ANGLE C IN DEG"
:Input Q
If Q≥ 180
:Then
:Disp "ANGLE 0≤A≤180"
:Goto 2
:Else
:Q+2π/360→F
```

Graphing Power © Dale Seymour Publications

TI–81 (cont.)

```
:Disp "ANGLE B"
:Disp E
:End
```

TI-82 (cont.)

```
:√(A²+B²-2ABcosF)→C
:(cos⁻¹((A²+C²-B²)/2AC))*360/2π→E
:(cos⁻¹((B²+C²-A²)/2BC))*360/2π→D
:Disp "SIDE C", C
:Disp "ANGLE A", D
:Disp "ANGLE B", E
```

Equation Tutor

This program generates positive coefficients of a linear equation in standard form, then coaches the user through the solution process.

The TI-82 program allows for interactive solving of two-step linear equations. The second stage of solving will not accept fraction form or decimal approximations of repeating decimals. When the coefficient of X is something like 9, 6, 7, it is best to use the divide choice from the menu.

TI-81

```
:ClrHome
:Disp "SOLVING"
:Disp "AX+B=C"
:Disp "WRITE PROBLEM"
:Disp "PRESS ENTER"
:Pause
:ClrHome
:Ipart (Rand*9)+1→A
:Ipart (Rand*9)+1→B
:Ipart (Rand*9)+1→C
:Disp "A="
:Disp A
:Disp "B="
:Disp B
:Disp "C="
:Disp C
:Pause
:Disp "SUBSTITUTE A,B,C"
:Disp "ENTER TO CONTINUE"
:Pause
:ClrHome
:Disp "WHAT OPERATION SHOULD YOU DO
    FIRST?"
:Disp "1)+ 2)- 3)* 4)/"
```

TI-82

```
PROGRAM:TUTOR
:ClrHome
:ClrDraw
:Split
:GridOff
:AxesOff
:Text(1,1, "SOLVE EQUATIONS OF")
:Text(8,1, "FORM AX+B=C")
:Text(16,1, "PRESS ENTER FOR")
:Text(24,1, "FIRST EQUATION")
:Pause
:ClrHome
:ClrDraw
:iPart (rand*9)+1→A
:iPart (rand*9)+1→B
:iPart (rand*9)+1→C
:Text(1,1, A, "X+", B, " = ", C)
:Text(24,40, "PRESS ENTER FOR NEXT
    STEP")
:Pause
:Lbl 4
:Menu("OPERATION TO SOLVE", "ADD", G,
    "SUBTRACT", H, "MULTIPLY", I,
    "DIVIDE", I)
```

Graphing Power © Dale Seymour Publications

TI-81 (cont.)

```
:Input E
:If E≠2
:Goto 7
:Lbl 2
:Disp "WHAT NUMBER DO YOU SUBTRACT?"
:Input M
:If B≠M
:Goto 2
:C-B→D
:Disp "NOW EQUATION IS"
:Disp "AX=D"
:Disp "WITH D="
:Disp D
:Disp "ENTER TO CONTINUE"
:Pause
:Disp "WHAT OPERATION SHOULD YOU DO?"
:Disp "1)+ 2)- 3)* 4)/"
:Lbl 9
:Input K
:If K≠4
:Goto 1
:Lbl 8
:Disp "WHAT NUMBER DO YOU DIVIDE BY"
:Input J
:If J≠A
:Goto 4
:D/A→X
:Disp "X="
:Disp X
:End
:Lbl 1
:Disp "THINK AGAIN"
:Goto 9
:Lbl 7
:Disp "THINK OPPOSITE SIGN"
:Goto 3
:Lbl 4
:Disp "THINK AGAIN"
:Goto 8
```

TI-82 (cont.)

```
:Lbl G
:Input "WHAT VALUE", M
:If M=-B
:Goto 9
:Disp "TRY NEW VALUE"
:Goto G
:Lbl H
:Input "WHAT VALUE", M
:If M=B
:Then
:-M→M
:Goto 9
:Else
:Disp "CHECK YOUR VALUE"
:Goto G
:End
:Lbl I
:Disp "CHECK OPERATION"
:Pause
:Goto 4
:Lbl 9
:Text(7,11,M, "              ", M)
:Horizontal 7
:Text(13,11,A, "X=", M+C)
:Pause
:Lbl 7
:Menu("NEXT OPERATION", "ADD", T,
    "SUBTRACT", T, "MULTIPLY", U,
    "DIVIDE", V)
:Lbl T
:Disp "CHECK OPERATION"
:Pause
:Goto 7
:Lbl U
:Input "BY WHAT VALUE", N
:If A*N≠1
:Then
:Disp "CHECK VALUE"
:Goto U
:Else
:Text(19,6,A*N, "X=", (M+C)*N)
:End
:Pause
```

TI-82 (cont.)

```
:Lbl V
:Input "BY WHAT VALUE", N
:If A/N≠1
:Then
:Disp "CHECK VALUE"
:Goto V
:Else
:Text(19,6,A/N, "X=", (M+C)/N)
:End
```